I0877968

In The Thickest of The Fray

Mississippians At Gettysburg In Their Own Words

Also by J. Douglas Ashton:

William Barksdale, CSA: A Biography of the United States Congressman and Confederate Brigadier General

Also by Joseph L. Owen:

A Fine Introduction to Battle: Hood's Texas Brigade at The Battle of Eltham's Landing, May 7, 1862

Historical Reminiscences of the Civil War: Sgt. J.B. Polley, Hood's Texas Brigade

Texans at Gettysburg: Blood and Glory with Hood's Texas Brigade

Unceasing Fury: Texans at the Battle of Chickamauga, September 18-20, 1863

Texans at Antietam: A Terrible Clash of Arms, September 16-17, 1862

A Texas Sergeant at Gettysburg: The Letters of Sgt. Val C. Giles, 4th Texas Infantry

A Texan First: Color Sgt. George A. Branard of the 1st Texas Recalls the Civil War

Lone Star Valor: Texans of the Blue & Gray at Gettysburg

In The Thickest of The Fray

Mississippians At Gettysburg In Their Own Words

Joseph L. Owen & J. Douglas Ashton

FOX RUN PUBLISHING
QUALITY PUBLISHING ONE BOOK AT A TIME

© 2024 by Joseph L. Owen and J. Douglas Ashton

All rights reserved. No part of this publication may be reproduced, stored in a retrieval system, or transmitted, in any form or by any means, electronic, mechanical, photocopying, recording, or otherwise, without the prior written permission of the publisher.

Publisher's Cataloging-in-Publication Data
provided by Five Rainbows Cataloging Services

Names: Owen, Joseph L., author. | Ashton, Douglas, author.
Title: In the thickest of the fray : Mississippians at Gettysburg in their own words / Joseph L. Owen [and] J. Douglas Ashton.
Description: Winston Salem, NC : Fox Run Publishing, 2024. | Includes bibliographical references and index.
Identifiers: LCCN 2024950840 (print) | ISBN 978-1-945602-27-6 (hardcover) | ISBN 978-1-945602-28-3 (paperback)
Subjects: LCSH: Mississippi--History--Civil War, 1861-1865--Campaigns. | Gettysburg, Battle of, Gettysburg, Pa., 1863. | Pennsylvania--History--Civil War, 1861-1865--Campaigns. | United States--History--Civil War, 1861-1865--Campaigns. | Nationalism--Confederate States of America. | BISAC: HISTORY / United States / Civil War Period (1850-1877) | HISTORY / Military / United States. | HISTORY / United States / State & Local / South (AL, AR, FL, GA, KY, LA, MS, NC, SC, TN, VA, WV) | HISTORY / United States / State & Local / Middle Atlantic (DC, DE, MD, NJ, NY, PA)
Classification: LCC E516 .O94 2024 (print) | DDC 973.7/1--dc23.

Cover design by Sandra Miller

Cover images:

Mississippi Monument photo, Courtesy: randy-steele.pixels.com under license

Terrific Fighting at the Angle, The Peach Orchard by Franklin Dullin Briscoe

Library of Congress Control Number (LCCN): 2024950840

Published by
Fox Run Publishing LLC

http://www.foxrunpub.com/

Dedication

This book is respectfully dedicated to the enduring memory of the brave sons of the Magnolia State within the brigades of Barksdale, Davis, and Posey who fought so valiantly and sacrificed so much during the three days of battle at Gettysburg.

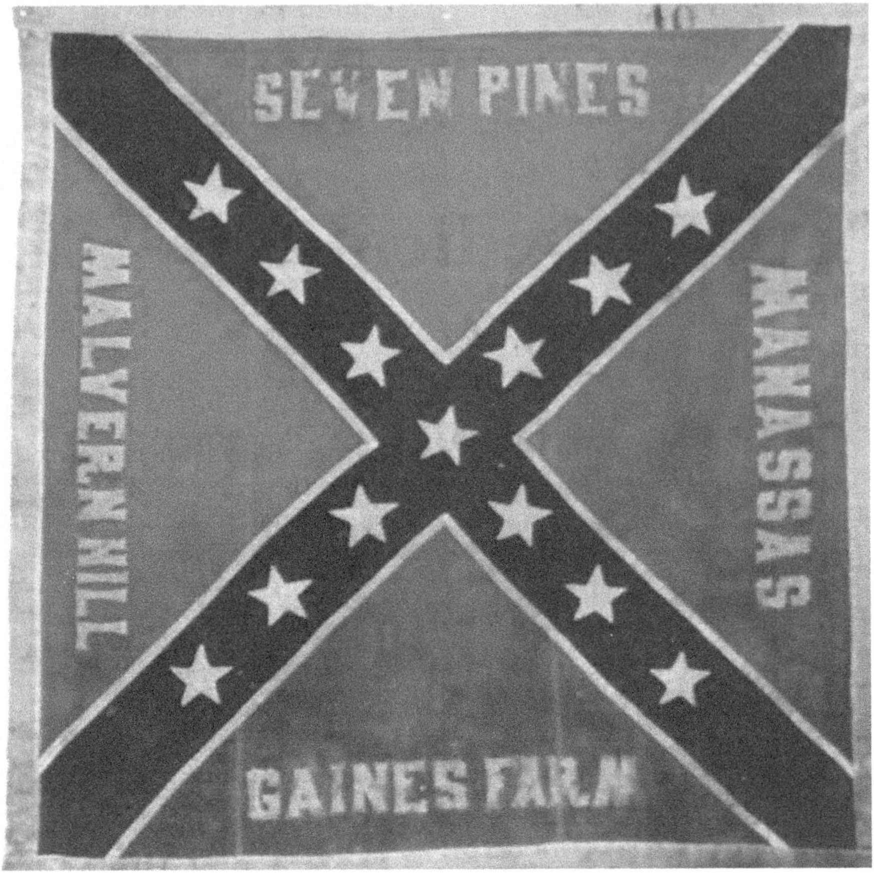

2nd Mississippi Infantry battle flag

(Mississippi Department of Archives and History)

The history of Gettysburg
with no mention of Mississippi,
is the story of the Alamo without Crockett,
Thermopylae without Leonidas

A Mississippi Vet. quoted in *The Weekly Democrat-Times*
(Greenville, Mississippi), June 21, 1902

13th Mississippi Infantry Battle Flag

(Jess N. McLean, Miss. Dept. of Archives and History)

16th Mississippi Infantry battle flag

(American Civil War Museum)

Table of Contents

List of Images

List of Maps

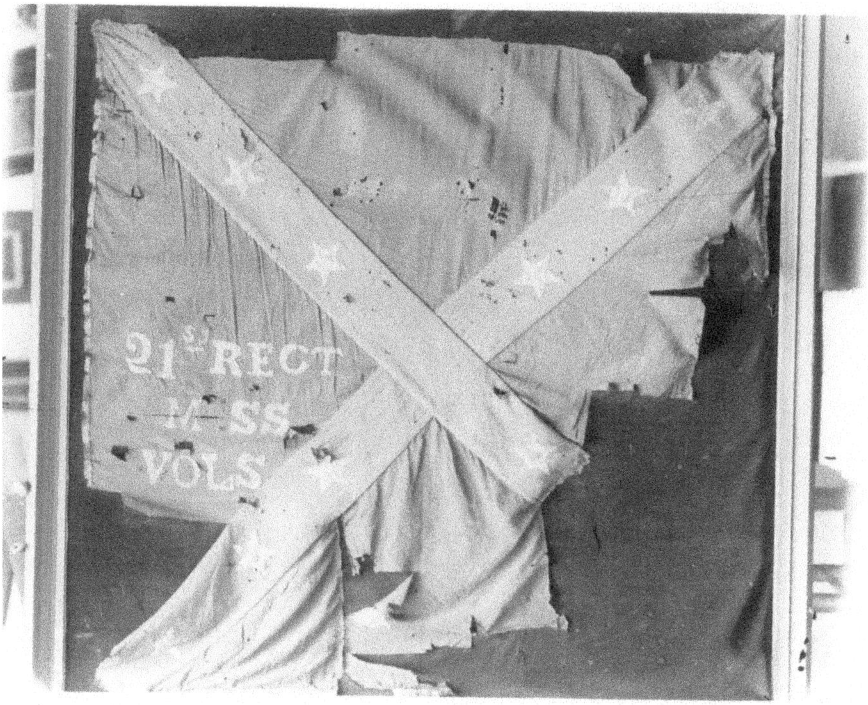

21st Mississippi Infantry battle flag

(Old Statehouse Museum, Vicksburg)

Surviving fragments of 42nd Mississippi Infantry battle flag

(Mississippi Department of Archives and History)

Foreword

Is Mississippi ready for this day, General?
– Robert E. Lee (Martin Sheen) in the movie Gettysburg

When I was a child, in the Old South tradition, my Mother visited often with friends and relatives about Houston, Mississippi. We often called upon Mary Thornton Hickman, my great-grandmother's sister's daughter. A bona fide member of the Daughters of the American Revolution (D.A.R.) and a First Family of Virginia (F.F.V.) member. Cousin Mary embodied a living, breathing "Miss Daisy" except she did the driving in her 1964 Chevrolet Bel Air (whose car is now "resting" in my Cashtown, Pennsylvania yard). It's all convoluted in the South sometimes.

Cousin Mary regaled me with tales of Southern ancestors, long gone, whose presence on Earth lay written in the family Bible. She related the story of the three brothers from the "illustrious" Hill family line that had served in the "The War." Randolph served in the 24th Mississippi and died a "glorious" death from dysentery in 1862 at Corinth. The second son fell at Gettysburg. "Thomas W. Hill, 11th MS, Co. H, Chickasaw Guards, died July 1 or 3, 1863, Gettysburg, PA," states the family Bible. He lost his life in Pickett's Charge (The secret to how I figured out the exact death date is contained in this book). He died for a "righteous cause" states the Mississippi Monument. It's all convoluted in the South sometimes.

With two of their three sons dead, Mississippi came for the Hill's last boy who had come of age in 1863. Words are inadequate to describe the anguish the family must have felt. In 1864, over the objections of his parents, William Lafayette Hill enlisted in the 6th Mississippi Cavalry and served in the command of Nathan Bedford Forrest. William's parents insisted that his "body servant," a slave named Dave Hill, accompany him to "The War." The two young men had grown up together and were reportedly close. Even after freedom came for Dave, they remained friends, visiting regularly. William is my great-great grandfather. Cousin Mary knew both of them. It's all convoluted in the South sometimes.

Besides blood, the one thread that united these brothers, these sons, these husbands, these friends, was their homeland, Mississippi - always Mississippi.

In The Thickest of The Fray is Doug Ashton and Joe Owen's first joint publication. The two authors bring plenty of experience to the subject. Joe has written and annotated several studies and memoirs on the Confederates from the Lone Star State: *Texans at Gettysburg*; *Unceasing Fury: Texans at the Battle of Chickamauga*; *A Fine Introduction to Battle: Hood's Texas Brigade at the Battle of Eltham Landing*; and *A Texas Sergeant at Gettysburg: The Letters of Sgt. Val C. Giles, 4th Texas Infantry*. He is also well versed on L.B.J. Doug, a lifelong student of the War of the Rebellion, is the author of *William Barksdale, CSA: A Biography of the United States Congressman and Confederate Brigadier Genera*l as well as several published articles in Civil War magazines.

In The Thickest of The Fray takes primary sources - letters, diaries, newspaper accounts, reminiscences - some published for the first time, to weave the story of Mississippi at Gettysburg. This book is not intended as a comprehensive transcription of every primary source. In each chapter on the battle, the authors "set the stage" first by giving an overview of the events and actions followed by the participant's accounts. In the following pages, the reader will get a glimpse of the common soldier, poor whites, mostly farmers, malnourished and underequipped, who nevertheless became some of the finest combat troops in America's history.

Follow Brig. Gen. Joe Davis's Brigade as they surge on the thrill of victory only to fall prey, seconds later, to a counterattack in the Railroad Cut. Hundreds of Mississippians were killed, wounded, and captured in the debacle. In the aftermath, Dr. Legrand Wilson described the battlefield as, "horrible beyond description. If every human being could have witnessed the result of the mad passions of men as I saw that night, war would cease, and there would never be another battle."

Feel the pounding of the horse's hooves as Brig. Gen. William Barksdale rides in front of his Mississippi Brigade. "The line before you must be broken," he exhorted his regimental commanders earlier, "to do so let every officer and man animate his comrades by his personal presence in the front line." With his hat off and hair flowing, Barksdale reminded one soldier of the "white plume of Navarre." Reigning up in the center of the brigade, Barksdale shouted, "Attention, Mississippians! Battalions, Forward!" One Union veteran commented afterwards, "it was the grandest charge that was ever seen by mortal man.... You just came on and on...." Mortally wounded inside Union lines, Barksdale assured his captors he had never "regretted the steps I have taken and although dying I do not regret my steps now." His last prayer asked God to "watch over and care for my dear wife and my boys, may God be a father to them, tell them to be good men and brave, always defend the right."

Hug Mother Earth, like mortal men did on July 3 waiting to charge, as Union shells ricochet and explode along Seminary Ridge. Look and you might see a stretcher bearing Jeremiah Gage, his body torn apart by shrapnel. "Remember that I am true to my country and my greatest regret at dying is that she is not free," he scribbled in his last letter home to Mother. Forty-five minutes later it was all over. "Upon the receding wave of the great charge, came a heavy drift of shattered humanity," wrote Dr. Joseph Holt. Thomas Hill was not among them. Like the fate of so many Mississippians, Tom probably lies in an unmarked grave in Hollywood Cemetery at Richmond, Virginia.

These were the men who propelled their Country to the cusp of Independence. In the end, all they really wanted was to go home – to Mississippi. Many never made it. I hope you enjoy meeting my heroes.

"Mississippi is ready, General!"

Matt Atkinson
Cashtown, Pennsylvania

48th Mississippi Infantry battle flag

(Mississippi Department of Archives and History)

Acknowledgements

We would like to thank Keith Jones and Eric Wittenberg for their outstanding support and for believing that the stories, articles and letters of the brave Mississippians that fought at Gettysburg should be told. Matt Atkinson for writing the Foreword — his advice and knowledge of the Mississippi regiments and the soldiers who fought at Gettysburg were invaluable. We are also grateful to retired Gettysburg National Military Park Historian John Heiser. His knowledge, support, and formal endorsement are deeply appreciated. Sincere thanks are extended to prolific Mississippi historian, author, and publisher Grady Howell for his review of our manuscript, formal endorsement, and permission to reproduce soldier photos from For Dixie Land I'll Take My Stand! Vol. IV . Thanks go to eminent historian and author Scott Mingus for reviewing our manuscript and formally endorsing it. Helpful advice from Gettysburg Licensed Military Park Battlefield Guides Rob Abbott and Jim Hessler is deeply appreciated. Sincere thanks are also extended to the Mississippi Department of Archives and History for allowing us to use the letters, photographs of the soldiers and flags that were at the Battle of Gettysburg; University of Mississippi Archives; Jordan Rushing of the Vicksburg, Mississippi's Old Courthouse Museum; Texas State Preservation Board; Baylor University; Library of Congress: John Hoopes for allowing us to use the photographs and accounts of his ancestor, Private William M. Abernathy of the 17th Mississippi Infantry; Jess McLean for his exhaustive research culminating in the valuable resource, The Official Records of the 13th Mississippi Infantry Regiment, and permission to use his photo of their battle flag. Thanks also to native Mississippian and author Jeff Giambrone for his continuing efforts to preserve the memory of these Mississippi soldiers. Finally, kudos to cartographer Hal Jespersen for his outstanding maps to clarify the positions, movements, and battle actions of the Mississippi regiments during the Gettysburg campaign.

Joe Owen / Doug Ashton

July18, 2024

I would like to thank friend and co-author Doug Ashton for joining me in telling the story of the Mississippi soldiers and regiments. Without Doug's knowledge, narrative, and organization, this book would never have been completed. Thank you Doug!

The support and enthusiasm of my parents Tom and Jean Owen, and brothers Luke Owen and Adam Owen was very important.

My late wife Cathy. Throughout the wonderful years that we had together, her love, support, advice, and understanding means more than I can possibly say. I Love you Cathy

Joe Owen

July 18, 2024

I am indebted to friend and co-author Joe Owen for asking me to come on board for this project. It has been a great experience and a logical progression from my earlier efforts on William Barksdale's first definitive biography. I trust that, together, we have done justice to the memory of the brave Mississippians who sacrificed so much during three momentous days of battle at a crossroads town in southern Pennsylvania — so far away from home and family.

As always the love and support of my loving wife Susie, and daughters Laura & Amy, mean so much to me.

Doug Ashton

July 18, 2024

Mississippi Governor John J. Pettus

Mississippi in the Confederacy

Mississippi—The Magnolia State—was the second state to join the Confederacy. It formally seceded from the Union on January 9, 1861, with the passing of the Ordinance of Secession at a meeting of the Secession Convention at the state Capitol in Jackson, Mississippi. The vote was 84-15 in favor and an eyewitness recorded that "Even before the Ordinance was passed, artillery fire could be heard, firing salutes in honor of the position the state was assuming." A formal two-hour signing ceremony followed on January 15, 1861 in the presence of Governor John J. Pettus, State Legislature members, state officials and spectators.[1]

Arguably no Southern state contributed more to the Confederacy on a per capita basis —nor suffered greater economic loss—than Mississippi. Admitted to the Union in 1817 during the Great Migration westward, its fertile soil and ideal climate propelled Mississippi to the forefront of Southern cotton production during the first half of the nineteenth century. By 1860, King Cotton had rendered Mississippi one of the wealthiest states in the Union. Huge fortunes had been made and "the many millionaires and magnificent mansions of Natchez were internationally known [with] land subject to taxes valued at $143,000,000." Slaves valued at $218,000,000 increased personal wealth to approximately $361,000,000—tenth highest state within the Union. However, by 1865 four years of civil war had totally devastated Mississippi making it one of the poorest—if not the poorest! The fate of Jackson, the state capital, is a case in point. On May 14, 1863 Union forces under Major General William T. Sherman captured and occupied the city. Following his victory at Vicksburg, Sherman returned and laid siege to Jackson during the period of July 10-16, 1863 virtually destroying it. Henceforth it was nicknamed "Chimneyville". On July 18, 1863 Sherman reported to Grant: "Jackson will no longer be a point of danger. The land is devastated for 30 miles around." Indeed, pre-eminent Mississippi historian John K. Bettersworth

1. Mississippi Commission On The War Between The States, *Mississippi in the War between the States: Civil War Centennial* (Jackson, MS, 1960), 5.

observed that in Mississippi "a struggle that began as the breaking of a nation ended with the breaking of a state."[2]

In 1860 Mississippi's free white population was 354,674 compared with a slave population of 436, 631— the latter was the highest within the Confederate states. It is estimated that 49% of the free families owned slaves, also highest within the Confederacy, followed closely by South Carolina at 45%. Slave owning families within the remaining Confederate states ranged between 37% and 20%. Mississippi furnished the Confederacy with a total of 85,000 troops representing 24% of the free population topping all other sister states within the Confederacy. Georgia posted the second highest recruitment rate at 21.8%. (See Appendix A)[3]

Mississippi also provided the Confederate States of America with its first and only president—Jefferson Davis. Although born in Kentucky (as was Abraham Lincoln) Davis relocated to Mississippi at an early age and operated a large cotton plantation, Brierfield, near Davis Bend. Davis launched his political career here and served his adopted state ultimately representing Mississippi as United States Senator in Washington and finally became Secretary of War under President Franklin Pierce.

Mississippi supplied the Confederate cause with thirty generals. Although none ascended to full general, one became a lieutenant general, five became major generals, and twenty-four led as brigadier generals as follows:

Lieutenant General: Stephen D. Lee

Major General: Samuel G. French, William T. Martin, Earl Van Dorn (murdered), Edward C. Walthall, William H. C. Whiting (mw)

Brigadier General: Wirt Adams, William E. Baldwin (accidental death), William Barksdale (mw), Samuel Benton (mw), William L. Brandon, William F. Brantley, James R. Chalmers, Charles Clark, Douglas H. Cooper, Joseph R. Davis, Winfield S. Featherston, Samuel W. Ferguson, John W. Frazer, Samuel J. Ghoulson, Richard Griffith (mw), Nathaniel H. Harris, Benjamin G. Humphreys, Mark P. Lowrey, Robert Lowry, Carnot Posey (mw), Claudius W. Sears, Jacob D. Sharp, Peter B. Starke, William F. Tucker

mw (mortally wounded)

---

2. Ibid., 14, 35; Jeff T. Giambrone, *Remembering Mississippi's Confederates* (Charleston, SC,2012), 7-8; John K. Bettersworth, ed., *Mississippi in the Confederacy: As They Saw It* (New York, 1970), xxxi
3. *The Civil War Almanac: A Book of Compelling Facts, Figures, and Expert Opinions*, Civil War Monitor (Kettering, Ohio, 2018), 12-13, 16-17, 20-21. Numbers are based upon 1860 Census.

The 85,000 officers and enlisted men hailing from Mississippi fought and died on numerous battlefields both within Mississippi and in every far-flung theater of war. Mississippi's strategic location bordering the mighty Mississippi River—the Father of Waters—was recognized by the military and political leadership of both sides resulting in dozens of battles and skirmishes at or near the Magnolia State's major cities such as Corinth, Iuka, Vicksburg (campaign towns include: Chickasaw Bayou, Grand Gulf, Port Gibson, Raymond, Jackson, Champion Hill), Brice's Crossroads, and Tupelo. In the western theater Mississippians fought valiantly at the battles of Shiloh, Stones River, Chickamauga, Atlanta, Franklin, Nashville, and Bentonville. While in the eastern theater Mississippians distinguished themselves at First Manassas, Leesburg, Seven Days Battles, Second Manassas, Sharpsburg, Fredericksburg, Chancellorsville/Second Fredericksburg, Gettysburg, Wilderness, and the Overland Campaign, right up to Appomattox Courthouse.

Following is the breakdown of Mississippi's recruitment efforts as compiled by Stewart Sifakis in his *Compendium Of The Confederate Armies: Mississippi*:

CAVALRY: 80 Battalions/Regiments

ARTILLERY: 44 Batteries including Independents

INFANTRY: 93 Battalions/Regiments[4]

The majority of Mississippi's forces served inside the expansive geographic area comprising the western theater as part of either The Army of the Department of Mississippi and East Louisiana, or the ill-fated Army of Tennessee. In early 1864 this latter army was reinforced with a heavy infusion of Mississippians who had been defending their native state resulting in the heaviest concentration of Mississippi troops ever—33 regiments/battalions![5]

Beginning with the Battle of First Manassas in July 1861 roughly 16,000 Mississippians saw Confederate service in Virginia during the remainder of the war as part of the Army of Northern Virginia under General Robert E. Lee. The initial five regiments who fought at Manassas were the 2nd Mississippi, 11th Mississippi, 13th Mississippi, 17th Mississippi, and 18th Mississippi Infantry regiments. By year-end 1861 another five Mississippi units were transferred to Virginia—namely the 12th Mississippi, 16th Mississippi, 19th Mississippi, and

4. Stewart Sifakis, *Compendium of the Confederate Armies: Mississippi* (Westminster, Md: Heritage Books, Inc, 2007).
5. *Remembering Mississippi's Confederates*, 53; *Mississippi in the War between the States: Civil War Centennial*

21st Mississippi Infantry Regiments along with the 2nd Mississippi Infantry Battalion. The 42nd Mississippi Infantry Regiment followed in summer 1862, and finally in 1864 the 26th Mississippi Infantry Regiment became the final regiment to deploy in Virginia. In addition, one cavalry regiment—The Jeff Davis Legion—and two artillery batteries—the Madison Light Artillery and the Confederate Guards Artillery—served in the Army of Northern Virginia. Only 698 of these Mississippians were left to surrender at Appomattox![6]

A post-war study by John L. Power, Mississippi's former superintendent of state army records, concluded that the attrition rate for Mississippi's soldiers was 30 per cent—12,000 killed or died of wounds, and 15,000 perished from disease![7] Power also estimated that an additional 6000 were lost to desertion as fears for the safety of loved ones back in Mississippi gnawed at these Mississippi soldiers serving in far-flung locales later in the war. Indeed, The Magnolia State paid an enormous sacrifice upon the altar of the Confederate cause for independence.

6. Ibid., 31.
7. Ibid., 7.

Mississippi Brigades Present at Gettysburg

Three brigades of Mississippians comprising eleven regiments totaling 5,247 officers and men fought at Gettysburg during the three-day battle—engaging in part on each day. Total losses were 1,948 (37%) consisting of 460 (9%) killed; 1,227 (23%) wounded; 261 (5%) missing.

Percentage losses were heaviest within William Barksdale's Mississippi Brigade in Lafayette McLaws's Division of James Longstreet's First Corps who fought at the Peach Orchard on the second day, i.e. total loss 804 (50%) consisting of 156 (10%) killed; 470 (29%) wounded; 178 (11%) missing. Next highest casualty rate was in Joseph R. Davis's Mississippi Brigade in Henry Heth's Division of A.P. Hill's Third Corps portions of which fought on both the first and third days of the battle in the vicinity of the Railroad Cut and in the Pickett-Pettigrew Assault, i.e. total loss 1,032 (45%) consisting of 289 (13%) killed; 677 (29%) wounded; 66 (3%) missing. Lowest casualties were inflicted upon Carnot Posey's Mississippi Brigade of Richard H. Anderson's Division of A.P. Hill's Third Corps which was only partially engaged on the second day near the Bliss farm, i.e. total loss 112 (8%) consisting of 15 (1%) killed; 80 (6%) wounded; 17 (1%) missing.

Barksdale's Mississippi Brigade

Longstreet's First Corps
McLaws' Division
Brig. Gen. William Barksdale Commanding
13th Mississippi - Col. James W. Carter (kia)
17th Mississippi - Col. William D. Holder (w)
18th Mississippi - Lt. Col. Thomas M. Griffin (w)
21st Mississippi - Col. Benjamin G. Humphreys

William Barksdale, although born in Tennessee, emigrated to Mississippi at age eighteen with his three brothers. A skilled lawyer and newspaper editor, Barksdale was elected to four consecutive terms in Congress at Washington where he became an ardent defender of slavery and states' rights. He served in the Second Mississippi Rifles

Brigadier General William Barksdale

(National Archives and Records Administration)

during the Mexican War as assistant commissary with the rank of captain but saw no actual combat. When Mississippi seceded Barksdale was appointed Quartermaster General of the Magnolia State but soon resigned and was elected colonel of the 13th Mississippi. Despite no formal military training Barksdale possessed a remarkable talent for leadership and self-learning and, with diligence, quickly acquired military know-how and tactical skills.

This regiment saw service at First Manassas where they helped to rout Union forces on the Confederate left flank. At the Battle of Ball's Bluff (Leesburg) the 13th Mississippi protected the Confederate rear by repelling a vastly superior enemy force at Edwards Ferry. During the army reorganization of 1862 the Mississippi Brigade was formed and put under command of Brig. Gen. Richard Griffith and saw active service during the Seven Days' battles. When Griffith was mortally wounded at Savage Station, Barksdale was promoted to command the brigade. At Malvern Hill he led the Mississippi Brigade from the front and personally advanced the colors under severe enemy fire—an astounding act of courage which won praise from General Robert E. Lee himself! Fighting side by side with Brig. Gen. Joseph B. Kershaw's South Carolina Brigade, Barksdale and his Mississippians stampeded Union forces off Maryland Heights which commanded Harpers Ferry. Posting of Confederate artillery on the heights resulted in the surrender of the Harpers Ferry garrison. After a grueling forced march to Sharpsburg, which diminished the brigade's strength by half, Barksdale's Mississippi Brigade nevertheless spear-headed the Confederate counterattack in the West Woods that decimated Maj. Gen. John Sedgwick's Union division at the battle of Antietam (Sharpsburg). At the battle of Fredericksburg Barksdale's Mississippi "sharpshooters" stoutly defended against the laying of Union pontoon bridges and subsequently continued their defense through the streets of the town. These actions were both military firsts and gained the admiration of the Confederate forces who witnessed these heroic actions. During the subsequent Chancellorsville Campaign Barksdale's Mississippi Brigade was chosen by Lee himself as part of the minimal force left in place at Fredericksburg to defend the rear of the Army of Northern Virginia against the vastly superior Union forces of Sedgwick's VI Corps. With their overwhelming numbers the Federals pierced Barksdale's thin line defending Marye's Heights causing the Mississippi Brigade to retreat for the first time ever! Despite this setback, Confederate forces soon rallied and managed to stem the Federal breakthrough. With their reputation tarnished, Barksdale and his Mississippians burned with a desire for redemption as they trudged toward Gettysburg.

Barksdale's Brigade was the most battle hardened of the three Mississippi brigades that fought at Gettysburg. By July 1863 Barksdale's veterans had earned the reputation as one of the hardest fighting and most disciplined brigades within the illustrious Army of Northern Virginia. Barksdale was just a few weeks shy of his forty-second birthday when he was mortally wounded at Gettysburg on July 2nd.

Posey's Mississippi Brigade

A.P. Hill's Third Corps
Anderson's Division
Brig. Gen. Carnot Posey Commanding
12th Mississippi - Lt. Col. Merry B. Harris
16th Mississippi - Col. Samuel E. Baker
19th Mississippi - Col. Nathaniel H. Harris
48th Mississippi - Col. Joseph M. Jayne

Carnot Posey was a native Mississippian from Woodville in Wilkinson County. Posey was an accomplished lawyer and planter and he later served as U.S. District Attorney in Mississippi. Posey saw active military service as a first lieutenant under Colonel Jefferson Davis in the First Mississippi Volunteer Regiment which gained fame as the Mississippi Rifles during the War with Mexico. Notably, Posey was wounded during the Battle of Buena Vista in February 1847. Upon Mississippi's secession Posey recruited the Wilkinson Rifles serving as their captain until they mustered in as Company K in the 16th Mississippi Regiment with Posey elected colonel. The 16th Mississippi fought under Major General "Stonewall" Jackson during his famous Shenandoah Valley Campaign in 1862. During the battle of Cross Keys Posey was wounded in the chest and arm, incapacitating him for two months. Although not fully recovered he finally rejoined the regiment in time to participate in the Confederate victory over General John Pope's Army of Virginia at Second Manassas.

Although the remaining three regiments of what later became Posey's Mississippi Brigade were all recruited at the same time together with the 16th Mississippi, each was assigned to a different brigade. Unlike the 16th Mississippi the other three, i.e. the 12th , 19th, and 48th (formerly the 2nd Mississippi Battalion) Mississippi regiments were assigned to General Joseph E. Johnston's army in Virginia where they participated in the Battle of Seven Pines and during the Seven days battles at Gaines's Mill and Glendale. The 12th, 19th, and 48th were assigned to Brig. Gen. Winfield S. Featherston's Brigade just before the Seven Days and the 16th Mississippi was added prior to the Second Manassas Campaign. Due to Featherston's severe illness Posey

Brigadier General Carnot Posey

(Confederate Military History by Evans)

was put in temporary command of the brigade prior to the Maryland Campaign and led it through the capture of Harpers Ferry and the Battle of Sharpsburg. However, ordered to support the desperate Confederate defense along the Sunken Road, when the division commander Richard H. Anderson was severely wounded early in the action, organization disintegrated, and resistance evaporated. Later, when Featherston was transferred to Mississippi, Posey was finally promoted to brigadier general formally assuming command of the brigade on January 19, 1863. At Chancellorsville, Anderson reported that "Posey's Brigade gallantly maintained its position against great odds and checked the further advance of the enemy."[1] Subsequently Posey's Mississippians performed only in a supporting role at Salem Church.

Although bloodied in combat Posey's Mississippians had not achieved the fighting reputation of Barksdale's men, nor had Posey accumulated Barksdale's level of experience in leading a full brigade into battle.

Davis's Mississippi Brigade

A.P. Hill's III Corps
Heth's Division
Brig. Gen. Joseph R. Davis Commanding
2nd Miss. - Col. J. M. Stone (w)
11th Miss. - Col. Francis M. Green (w)
42nd Miss. - Col. Hugh R Miller (kia)

Note: The 55th North Carolina Regiment was brigaded under Davis, but their numbers and casualty figures have been excluded.

For a breakdown of each Mississippi brigade by regiment and company see Appendix B.

Davis' Brigade was the most recently organized, the least battle-hardened overall, and led by the "greenest" commander of the three Mississippi brigades present at Gettysburg. Joseph R. Davis, an accomplished young lawyer and state senator, was also the privileged nephew of Confederate President Jefferson Davis. Although elected lieutenant colonel of the 10th Mississippi Regiment at the outset, on May 20, 1861 36-year-old "Joe" Davis was ordered to serve as the President's aide-de-camp (essentially personal secretary). The closest Joe came to any fighting was when he accompanied his uncle on an official visit during the Battle of First Manassas. Described as "the easy going, genial, embodiment of the Southern gentleman,"[2] what Joe lacked in aggressiveness he made up for in ambition to secure a field command. In summer of 1862, President Davis nominated his nephew for promotion to brigadier general but the Confederate Senate, balking at this act of nepotism, rejected the nomination. However, due to political maneuvering the President prevailed, and his second attempt

1. *OR* 37/1, No. 867, p. 851.
2. John Douglas Ashton, *William Barksdale, CSA* (Jefferson, NC, 2021); Bradley M. Gottfried, *Brigades of Gettysburg*, (Cambridge, MA, 2002) 410-416; 588-592; 613-621; Dunbar Rowland, *Military History of Mississippi 1803-1898*, Grady Howell, ed., (Madison, MS, 2003) 50-116; 121-129.

was successful. Joe Davis was officially promoted on September 15, 1862, becoming the most inexperienced brigadier general in the Army of Northern Virginia!

Two of the four regiments in the brigade mirrored their commander's lack of battle experience. The 42nd Mississippi was organized in the spring of 1862 and numbered over one-thousand men. Two companies were sent to guard the fords of the Rappahannock river near Fredericksburg for a few days but were recalled to Richmond on November 22, 1862 and the entire regiment was then transferred to North Carolina. Gettysburg would represent their baptism of fire. The 55th N.C. was organized near Raleigh early in 1862. With the exception of a skirmish to capture and briefly occupy the tidewater town of Washington, N.C. on September 6, 1862, this regiment remained in their native state and saw no further action prior to Gettysburg after being brigaded with the three Mississippi regiments.

Fortunately, the two remaining regiments—the 2nd Mississippi and 11th Mississippi—were both veteran contingents having seen heavy action at First Manassas, the Peninsula Campaign, the Seven Days Campaign (Seven Pines, Gaines's Mill and Malvern Hill), Second Manassas, and Sharpsburg. However, they were a study in contrast—the 2nd Mississippi was comprised of common farm boys and mechanics from a poor region of southern Mississippi while the 11th Mississippi hailed from the older and more affluent counties in northern Mississippi and consisted of dashing young students, clerks, and professionals. The University Greys (Company A) and the Lamar Rifles (Company G) of the 11th Mississippi gained renown since their ranks were filled with students/graduates of the University of Mississippi at Oxford. The reputation of the University Greys was cemented on the third day at Gettysburg when the company suffered 100% casualties during the Pickett-Pettigrew Charge![3]

The Order of Battle for the Mississippi regiments engaged at Gettysburg is presented in Appendix B.

3. Ibid.

Brigadier General Joseph R. Davis

(USAMHI)

CHAPTER 1
The March Toward Gettysburg

Following the battle of Chancellorsville the Confederate and Union armies returned to their former positions at Fredericksburg staring at each other across the Rappahannock River. However, true to his aggressive nature, General Robert E. Lee did not remain idle. Stonewall Jackson's death at Guinea's Station on May 10th necessitated a major reorganization of the Army of Northern Virginia. For greater flexibility, from the former structure of two large corps of four divisions each, Lee quickly created three smaller corps of three divisions each. While Longstreet retained command of First Corps, Second Corps (Jackson's) was now commanded by Richard Stoddart Ewell, and the newly-created Third Corps fell under Ambrose Powell Hill, the latter both West Point graduates, career soldiers, Virginians, and newly promoted lieutenant generals. Unfortunately neither Ewell nor Hill had previously commanded such a large force. Moreover, while Longstreet retained three seasoned divisional commanders, only three of the six divisions in the remaining two corps had leaders who had previously commanded a division. In particular Hill's Third Corps was comprised of thirteen brigades from eight different states. Lack of familiarity between brigades, between commanders and staff officers, and between Ewell/Hill who had never served directly under Lee and were unaccustomed to his command style of issuing discretionary orders, represented a serious liability which would have dire consequences at Gettysburg. Indeed, renowned historian Douglas Southall Freeman considered this reorganization "one of the major tragedies of the Confederacy."[1]

Barksdale's Mississippi Brigade continued under command of division commander Major General Lafayette McLaws in Longstreet's First Corps while both Carnot Posey's Mississippi Brigade under divisional Major General Richard H. Anderson and Joseph R. Davis's Mississippi Brigade under divisional Major General Harry Heth both belonged to A. P. Hill's newly formed Third Corps.

---

1. Douglas Southall Freeman, "Why Was Gettysburg Lost," *Lee: The Soldier*, Gary Gallagher ed., (Lincoln, NB: 1987) 459-60.

Lt. Gen. James Longstreet

(Library of Congress)

Lee immediately sought and received support from Confederate President Jefferson Davis and Secretary of War James J. Seddon for an invasion into Pennsylvania and commenced the pullout from Fredericksburg less than one month after his victory at Chancellorsville. While Hill's Third Corps remained in defensive positions at Fredericksburg opposite Hooker's Union army to screen Confederate movements, Lee stealthily withdrew both Longstreet's

Major General Lafayette McLaws

(Library of Congress)

and Ewell's corps westward toward the Shenandoah Valley enroute to Pennsylvania. As noted author and historian James I. (Bud) Robertson has concluded: "Lee's plan of march into the North was a bit complicated, but sound. Longstreet would remain at Culpeper to give aid to Hill should Hooker lunge southward. Ewell, with a brigade from [Stuart's cavalry] paving the way, would move rapidly northward through the Shenandoah Valley, cross the Potomac, and advance through Maryland into Pennsylvania. Then Longstreet would also start northward. While Longstreet acted as a screen, Hill would move westward from Fredericksburg and follow in Ewell's footsteps.

Jefferson Davis, president of the Confederate States of America and uncle of Brigadier General Joseph R. Davis

(Library of Congress)

Longstreet's corps would bring up the rear as the whole army carried the war onto Northern soil."[2] Accordingly on June 2nd, with Barksdale's Brigade leading the march out of Fredericksburg along the Telegraph Road, McLaws' Division formed the vanguard of Longstreet's corps, with Hood's and Pickett's divisions following next day. Ewell's corps broke camp and followed on June 3rd while Hill's corps remained in position observing Hooker's forces until June 14th when, satisfied that Hooker's army had finally departed in search of Lee's army, Hill began his withdrawal. Anderson's division, including Posey's brigade, took the lead followed by Heth's division, including Joe Davis's brigade, with Pender's division following next day. Thus ironically Davis's Mississippians were among the last men of the Magnolia State to leave Fredericksburg but would be first to arrive on the battlefield of Gettysburg on July 1st.[3]

The route westward out of Fredericksburg passed through the Chancellorsville battlefield fought one month earlier. Several Mississippians recorded their ghastly observations in their diaries, letters, and accounts. Sergeant William H. Hill, 13th Mississippi, the twenty-two-year-old regimental quartermaster clerk recorded in his diary: "June 4, 1863 - Thursday, clear and very warm. The Brigade started at 6AM and marched to Raccoon Ford on the Rapidan River. We passed through Chancellorsville. *The country around is completely devastated for several miles and the stench from the dead bodies of men and horses is intolerable.* Marched 23 miles today."[4] Legrand J. Wilson, Assistant Surgeon of the 42nd Mississippi gave a more graphic account: "We marched through a portion of the battlefield of Chancellorsville, and witnessed some of the horrors of war. It had been about six weeks since that terrible battle was fought, *and the stench was, in places, perfectly awful.* The little hillocks scattered here and there along the route for several miles, were silent witnesses of the savagery of man, and the horrors of war. *These men had been buried hastily and many of them hardly covered beneath the sod, and here and there an arm or leg was sticking out, the flesh stripped from the bones by the beasts of the field and the vultures of the air that had congregated in countless numbers to satiate their hunger with human food.*" (emphasis added)[5]

The only Mississippians who saw action during the march into Pennsylvania were Barksdale's Mississippi Brigade who were called into

2. James I. Robertson, *General A.P. Hill: The Story of a Confederate Warrior* (New York, 1987) 200.
3. John Douglas Ashton, *William Barksdale, CSA* (Jefferson, NC, 2021) 204; Ibid.
4. Sgt. William Hill Diary, 1861-1863, MDAH, Z0217.00
5. Legrand James Wilson Memoir, 1902, *The Confederate Soldier*, reprint Bell Wiley ed. (Memphis, 1973), 110-11.

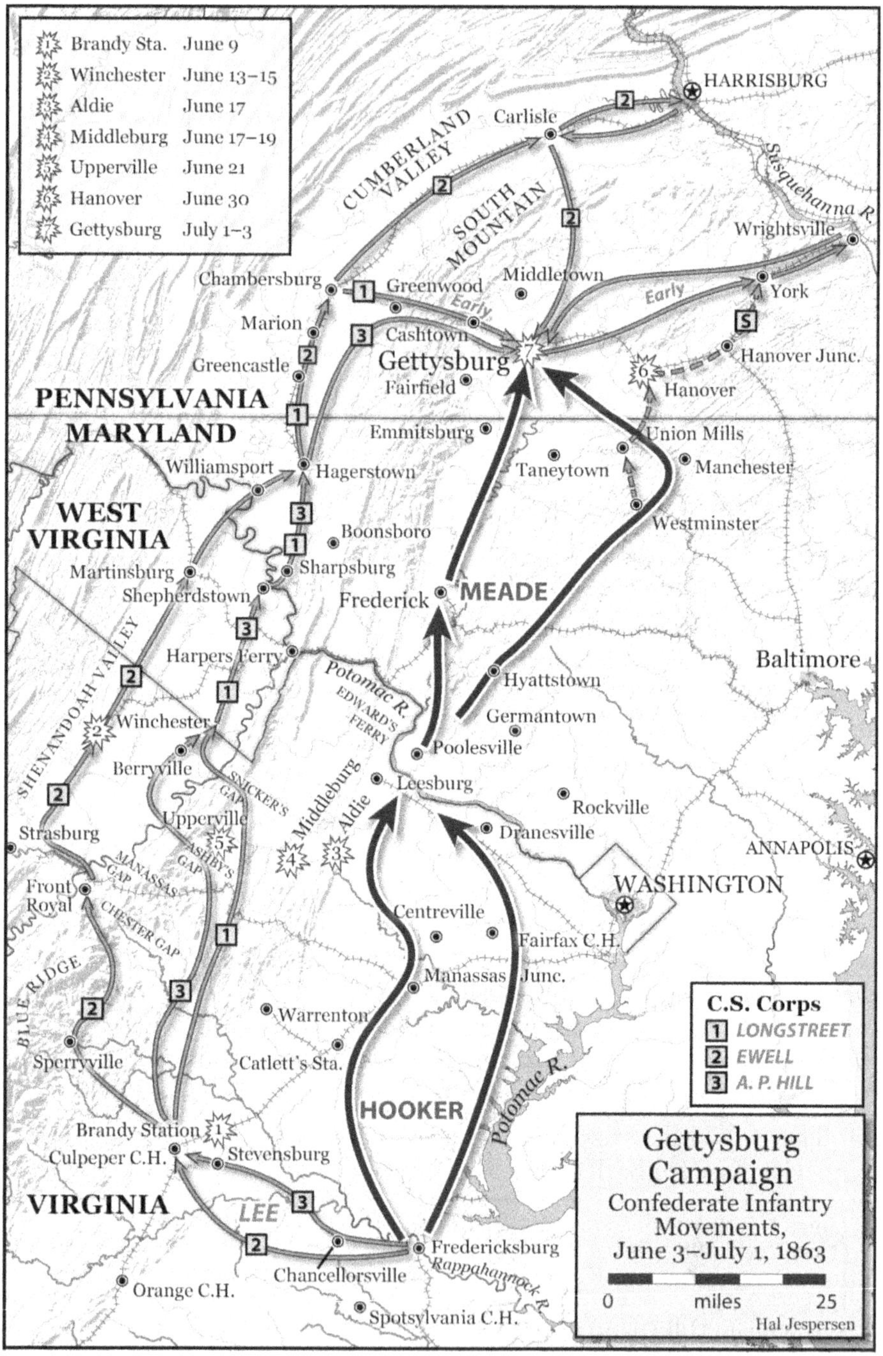
Brandy Sta. June 9
Winchester June 13–15
Aldie June 17
Middleburg June 17–19
Upperville June 21
Hanover June 30
Gettysburg July 1–3
HARRISBURG
Carlisle
CUMBERLAND VALLEY
SOUTH MOUNTAIN
Susquehanna R.
Wrightsville
York
Early
Middletown
Chambersburg
Greenwood
Marion
Cashtown
Gettysburg
Hanover Junc.
Hanover
Greencastle
Fairfield
PENNSYLVANIA
MARYLAND
Emmitsburg
Union Mills
Taneytown
Manchester
Williamsport
Hagerstown
WEST VIRGINIA
Boonsboro
Westminster
Martinsburg
Sharpsburg
Shepherdstown
Frederick
MEADE
Harpers Ferry
Baltimore
Potomac R.
EDWARD'S FERRY
Hyattstown
SHENANDOAH VALLEY
Winchester
Germantown
Poolesville
Berryville
SNICKER'S GAP
Middleburg
Aldie
Leesburg
Rockville
Upperville
Strasburg
ASHBY'S GAP
MANASSAS GAP
Dranesville
ANNAPOLIS
Front Royal
CHESTER GAP
WASHINGTON
Centreville
Fairfax C.H.
BLUE RIDGE
Manassas Junc.
Warrenton
C.S. Corps
LONGSTREET
EWELL
A. P. HILL
Catlett's Sta.
Sperryville
HOOKER
Potomac R.
Brandy Station
Culpeper C.H.
Stevensburg
VIRGINIA
LEE
Fredericksburg
Chancellorsville
Rappahannock R.
Orange C.H.
Spotsylvania C.H.
Gettysburg Campaign
Confederate Infantry Movements,
June 3–July 1, 1863
0 miles 25
Hal Jespersen

line of battle several miles east of Culpeper at Stevensburg to fend off an incursion by Federal cavalry. This occurred just prior to the battle of Brandy Station on June 9th which saw General J.E.B. Stuart's Confederate cavalry narrowly defeat Brigadier General Alfred Pleasanton's Union cavalry in the largest single cavalry engagement in the history of the Western hemisphere. Again on June 21st Barksdale's brigade was ordered to defend Ashby's Gap in support of Stuart against Union cavalry attacks. Otherwise the only other formidable enemies the Mississippi brigades had to face on the march were the forces of Mother Nature—oppressive heat and drenching rain!

Sergeant William Hill recorded in his diary for June 17th: "Wednesday, Clear and very warm. *The troops are suffering severely from the heat. Several of the men in the Division died today from being overheated.*" (emphasis added) In a letter to his wife dated June 21, 1863, Private Newton Nash, Co. I, 13th Mississippi wrote: "*The weather was intensely hot, as hot as I ever saw in Miss[issippi]. A great many fainted in the road with heat and several died from the same cause. I fainted on the way one day about two p.m. The Regt. had to go on and leave me and several others.* (emphasis added) Some of them said they would have to leave old mule behind this time. In about two hours it turned cool and commenced raining which revived me so much that I went on that night, caught up about dark, have been improving ever since."[6]

The next major event that virtually all the Mississippians commented upon in their letters and diaries was the crossing of the Potomac at either Boteler's Ford about one and a half miles below Shepherdstown, Virginia or another opposite Williamsport, Maryland. R.H. Anderson's Division crossed at Boteler's Ford with Posey's Mississippi Brigade wading across on the morning of June 24th. Heth's Division crossed the following day with Davis's Mississippi Brigade leading the way. Legrand J. Wilson of the 42nd Mississippi trailing the column and perched on the high ground 200 feet above the broad Potomac River described the scene:

> **The road turned squarely to the right and headed toward the river. The scene which presented itself was enough to arouse enthusiasm in every heart. Over the hills beyond the river as far as the eye could reach, a line of gray soldiers were marching, cheering as they went, their bright arms glittering in the sunshine. Below at the river the men were marching across holding their cartridge boxes to keep them dry. A short distance below the line**

6. Hill Diary, 1861-1863, MDAH Z0217.00; Nimrod Newton Nash letters, Old Courthouse Museum Collections, Vicksburg MS

> **of men, the artillery was crossing and above, the wagon train and ambulances. Bands were playing Dixie on the Virginia shore and Maryland, My Maryland beyond the river. The scene was grand and inspiring.**

Twenty-five- year-old First Lieutenant William H. Peel, Co. C, 11th Mississippi, a farmer from Chickasaw County, recorded his observations and touched upon the camaraderie within the ranks. He commented that the river was several hundred yards wide and about three feet deep. However, the strong current and sharp rocks made the crossing difficult for the struggling foot soldiers, cavalry, artillery and heavy wagons which elicited a wide array of yells and shouts. Occasionally some lost their footing and submerged as in a Baptist immersion, much to the delight of their companions who laughed uncontrollably and mocked them with insults of every kind.[7]

Amidst the revelry at least one Confederate soldier recognized the inevitable result of invasion. Private Wiley Heflin, Co. D., 11th Mississippi shouted out—"Boys, there are some of us crossing the river who will never recross again."[8]

McLaws' Division crossed at Williamsport on June 26th with Barksdale's Mississippians trailing. Barksdale's Brigade was therefore not only the final of the three Mississippi brigades to cross into Maryland but also the final infantry brigade in the Army of Northern Virginia to do so. For these Confederates the crossing was memorable for another unexpected reason. General McLaws later related that the Potomac here was quite deep and the wading across sufficiently difficult that his soldiers became very wet. A large quantity of whiskey had been confiscated in Williamsport so McLaws ordered that a "gill apiece [be given] to each man that wanted it." He proudly proclaimed, "in justice to my division I will assert that I never heard of anyone refusing it; the consequence was that the men were all in good humor," a fact confirmed by Private William Abernathy of Co. B, 17th Mississippi who described his fellow Mississippians as "about a lively set of men as ever tramped up that pike."[9]

For the members of the 2nd Mississippi and 11th Mississippi in Davis's Brigade the trek northward on the Hagerstown Turnpike through Maryland brought them close by Miller's cornfield on the

7. Ibid, 113; Steven H. Stubbs, *Duty*Honor*Valor: The Story of the Eleventh Mississippi Infantry Regiment* (Philadelphia, MS, 2000) 401-02; William H. Peel, Company C, Diary 1863-1865, Original Typescript Manuscript, MDAH.
8. Ibid., 402.
9. Major General Lafayette McLaws, "Gettysburg," SHSP, Vol. 7, 65; William Meshack Abernathy, "Confederate Memoir: Our Mess," *Confederate Veteran Magazine*, Vol. 2, 2003, 31-2.

Major General Richard Heron Anderson

(National Archives)

Sharpsburg battlefield where they had fought gallantly and been mowed down nine months earlier. The scene, even then, evoked painful memories as recorded by 1st Lt. William H. Peel:

> **Two hostile armies had met in dreadful conflict, the fatal consequences of which were legibly written on almost every surrounding object. The houses were pierced by numerous shells & shrapnel, the picket fences on either side of the road were perfectly riddled by rattling musketry, while many a little mound, scattered here & there over the field bespoke the final resting place of some friend or foe, the unfortunate victim of the horrible massacre of that eventful day.**[10]

10. Stubbs, *Duty *Honor *Valor*, 402.

General Ewell's Second Corps, far ahead in the vanguard, crossed the state line into Pennsylvania on June 26th and proceeded northward fanning out in several directions with the intent of capturing Harrisburg, the state capital. The Mississippians in Longstreet's and Hill's corps also commented upon crossing into Pennsylvania. Sergeant Hill of the 13th Mississippi made the following entry in his diary on June 27th:

> **We crossed the Pennsylvania line at 1 p.m. at the village of Middleburg. Passed through Greencastle, Pennsylvania at 3 p.m. The people are hiding their property and moving it out of our reach as fast as they can. But our Quartermasters and Commissaries have taken a large number of horses, mules, wagons, beef cattle, sheep, forage etc. They pay for these things where the people will receive our currency, and if they refuse give them a receipt to be paid at the end of the war. A large number of citizens have fled, leaving their homes unprotected.**

and next day added:

> **The people are much depressed and uneasy on account of our invasion of their state and express their dislike for us on all occasions. Our Army is living on the countryside and the Quartermasters are taking everything that the Army can use.**

To a man these Mississippi infantrymen were extremely impressed with the countryside and farms within Pennsylvania and freely recorded their observations. Private Newton Nash writing to his wife on June 28th stated: "[this] is the most beautiful country I ever beheld. I hate to see such a fine looking country in the possession of such people." Private Clendenen Black, Co. I, 13th Mississippi elaborated further: "This is the most beautiful country I ever saw, the houses are all fine, the barns are more magnificent than the dwelling houses."[11]

Lee issued specific orders that were read aloud to all his troops —"No private property shall be injured or destroyed," and posted sentries at private residences, restricted the soldiers to camp, and instructed them to treat the citizens courteously. General Carnot Posey confirmed, in a letter to his wife, that the measures were predominantly effective: "every effort is used to check any deprevations. No houses or barns are burnt, no women & children

11. Hill Diary, MDAH 1861-1863, MDAH Z0217.00; Nash Letters, Old Courthouse Museum, Vicksburg; Private Clendenen Black to Mother, June 28, 1863, in *Mississippi In The Confederacy: As They Saw It*, John K. Bettersworth ed. (Jackson, MS 1961) 163.

turned out of doors ... So you see we are not barbarians nor do we imitate the infamous example of our enemy."[12]

Private David Holt of the 16th Mississippi wrote of an encounter with a resident of Chambersburg which sheds some light on the situation:

> **The natives seemed kind but curious. They had a notion that we were a lot of ignorant savages, particularly we from Mississippi. Their surprise was great when they found us courteous and gentlemanly. Princes disguised in rags. Hoard and myself walked a short distance from camp one evening after roll-call and came to a cottage where a woman and her daughter were trying to get an unruly calf through a gate. We went to their assistance, and, after a healthy chase, got the calf in and the gate shut. Then the mother asked us what state we were from. When we said "Mississippi" she lifted up her hands and said, "Oh! I thought all persons from that state wore fierce-looking beards and carried bowie knives, but you two look mild enough. Can't we give you something to eat for your trouble?" We lost no time in accepting the invitation.**[13]

While major depredations and destruction were avoided and the soldiers well-behaved while under the direct discipline of their officers, invariably when left on their own, as Private Robert A. Moore, Co. G, 17th Mississippi recorded in his diary—"the souldiers are committing some depredations on private property," mainly in the form of foraging. Lieutenant Peel of the 11th MS noted the following in his diary:

> **We had not been many hours in bivouac, until turkeys, ducks, chickens, crocks of milk, butter, apple butter, & various other delicacies began to make their appearance in almost every mess, while the occasion[al] report of musket & the responsive squeal of a pig showed that even these unfortunate animals were not exempt from the cruelties of the retaliation we believed ourselves justified in inflicting.**[14]

12. Robertson, *General A.P. Hill*, 203.
13. David Holt, Thomas D. Cockrell, and Michael B. Ballard, *A Mississippi Rebel in the Army of Northern Virginia: The Civil War Memoirs of Private David Holt* (Baton Rouge: Louisiana State University Press, 1995), 192-3.
14. James W. Silver, A Life For The Confederacy: The War Diary of Robert A. Moore, Pvt. CSA (Jackson, TN: McCowat-Mercer Press Inc., 1959) 152; Stubbs, *Duty*Honor*Valor*, 405.

Jacob Hoke was a keen-eyed merchant with a store on the main street of Chambersburg. In his prolific eyewitness account, *The Great Invasion of 1863*, published in 1887, Hoke gave his detailed description of these Confederate invaders which can be considered representative of the general citizenry of Pennsylvania:

> **The Confederate infantry, as they marched through Chambersburg, presented a solid front. They came in close marching order, the different brigades, divisions, and corps, all within supporting distance of each other. Their dress consisted of nearly every imagineable color and style, the butternut predominating. Some had blue blouses, which they doubtless stripped from Union dead. Hats, or the skeletons of what had been hats, surmounted their partly covered heads. Many were ragged, shoeless, and filthy, affording unmistakable evidence that their wardrobe sadly needed to be replenished. They were however, all well armed and under perfect discipline. They seemed to move as one vast machine. Laughing, talking, singing, and cheering were not indulged in. Straggling was scarcely seen, but when some of them did wander from the lines, and caught any of our people in retired places, they did not hesitate to appropriate themselves hats, boots, watches, and pocketbooks. This proves that their good behavior when under the eyes of their officers was due to discipline rather than innate honesty and good breeding.**

Hoke voiced an inherent prejudice and marked dislike toward the common Confederate infantryman:

> **... while the rank and file of the infantry were uneducated, the lower classes, many of whom were conscripted into the service. Many of these seemed to have no intelligent understanding of the causes of the war, of what they were fighting for, or of the value of the Government they were undertaking to overthrow. When interrogated upon these points they would say they were fighting for** *their rights*. **What rights they had which they were in danger of losing, they did not attempt to tell. There was a marked difference in these men, which seemed to be a peculiarity of all from the same State.** ***Those from Mississippi and Texas were more vicious and defiant than those from other parts of the South.***[15]

15. Jacob Hoke, *The Great Invasion of 1863* (1887), reprint (New York, 1959) 208-09.

Despite its ragged appearance Lee's Army of Northern Virginia was extremely confident and spoiling for a fight. Captain Arthur J. Freemantle of Britain's famous Coldstream Guards, travelling with Longstreet's First Corps, observed that "the universal feeling in [Lee's] army was one of profound contempt for an enemy whom they have beaten so constantly, and under so many disadvantages." Private John S. McNeily of the 21st Mississippi later wrote of Barksdale's Mississippians at Gettysburg: "Never was a body of soldiers fuller of the spirit of fight, and confidence of victory." Colonel E. Porter Alexander, talented commander of the artillery reserve, opined that "[Lee] carried the best and largest army into Pennsylvania that he ever had in hand." Alexander concluded that this deepest penetration of the North now put them "somehow nearer the enemy's heart than we had ever been before. There seemed to be a prevalent feeling that fortune now favored us ...."[16]

One final exceptional role was to be played by a Mississippian which precipitated the monumental three-day battle at Gettysburg. As related by Major Moxley G. Sorrel, Longstreet's Chief of Staff, while Longstreet was on detached service in Suffolk prior to the Chancellorsville campaign,

> **a scout, more properly a spy, was placed at his service by the War Department. He was a man of about thirty years, calling himself a Mississippian, and was altogether an extraordinary character. he was paid in United States greenbacks. I approved requisition on the quartermaster every month for $150 for him. His time seemed to be passed about equally within our lines and the enemy's. Harrison (such was his name) always brought us true information. There was invariable confirmation of his reports afterwards... we had means, and did verify his account of himself as coming from Mississippi.**

As Longstreet prepared to leave Fredericksburg for the march into Pennsylvania, Sorrel continued his account:

> **Longstreet sent for his favorite scout Harrison. His instructions were to proceed into the enemy's lines, where he was to stay until the last part of June. Then he**

16. A.J.L. Freemantle., "The Gettysburg Campaign," *Two Witnesses at Gettysburg: The Personal Accounts of Whitelaw Reid and A.J.L. Freemantle*, Gary W. Gallagher ed. (St. James, NY, 1994) 119; J.S. McNeily, "Barksdale's Brigade at Gettysburg," Publications of the Mississippi Historical Society, Vol. XIV,1914, 235; E.P. Alexander, "Letter on Causes of Lee's Defeat at Gettysburg," *Lee The Soldier*, Gary W. Gallagher ed. (Lincoln, NB, 1996) 436; *Idem*, "Artillery Fighting At Gettysburg," *Battles & Leaders*, Vol. 3, 357-58.

Confederate spy, Henry Thomas Harrison

(H. Grady Howell, Jr.)

> **was to report to General Longstreet, it was hoped with the amplest and most accurate information...**

Sorrel continued his recollection of events:

> **At night [June 28th] I was aroused by a detail of the provost guard bringing up a suspicious prisoner. I knew**

> **him instantly, it was Harrison, the scout, filthy and ragged, showing some rough work and exposure. He had come to 'report to the general, who was sure to be with the army,' and truly his report was long and valuable.... Harrison gave us the first complete account of the operations of the enemy since Hooker left our front. He brought his report down to a day or two, and described how they were even then marching in great numbers in the direction of Gettysburg, with intention apparently of concentrating there. He also informed us of the removal of Hooker and the appointment of Major General George Gordon Meade to command of the Army of the Potomac.**

Recognizing the significance of these events, Sorrel immediately accompanied Harrison to see General Lee. Sorrel continued his account:

> **The General [Lee] heard him with great composure and minuteness. It was on this, the report of a single scout, in the absence of the cavalry, that the army moved.... The commanding general was not long in deciding. He sent orders to bring Ewell immediately back from the North about Harrisburg, and join his left. Then he started A.P. Hill off at sunrise for Gettysburg, followed by Longstreet. The enemy was there, and there our general would strike him. (emphasis added).**[17]

The greatest three-day battle in the Western Hemisphere was about to begin!

* * *

CONFEDERATE VETERAN MAGAZINE
MARCH, 1913

CORPORAL SAMUEL W. HANKINS
2ND MISSISSIPPI INFANTRY

Our next move was up the Rappahannock River near Fredericksburg, where we were attached to General Heth's division, A. P. Hill's corps, and did picket duty along the banks of the river until General Lee began his move the latter part of June into the enemy's country. On this move we forded the Potomac River. It was rather wide, but only about waist deep at the time. The water, however, was swift. The river bed was very rough and slippery, so that wading was

17. Brig. Gen. G. Moxley Sorrell, Bell Irvin Wiley ed., *CSA, Recollections of a Confederate Staff Officer*, (Wilmington, NC, reprint, 1995), 147-48, 152, 155-56.

Major General Henry Heth

(Library of Congress)

extremely difficult. Many got a ducking, to the great amusement of those who escaped. The artillery and wagons and those on horseback had to cross about a hundred yards below us. Some officer seated on his horse about midway the river was yelling at us to close up, when his horse stumbled and he went over the horse's head foremost into the water. An uproar by the soldiers followed his catastrophe. Our own brigade was the second to cross. Looking back after reaching the opposite side, we saw the army, miles in length, winding its way through the mountains like a huge snake down into the river. It was a grand sight.

Our first night in camp in Maryland was in the vicinity of Hagerstown. I was on provost duty that night, which was to keep the soldiers out of town. We made only a few arrests. The reason we had so little trouble was that General Lee had issued very stringent orders about molesting private property and committing depredations, and his men well understood him.

Hagerstown was strongly Union in sentiment, so all along the streets could be heard female voices singing Northern airs and war songs; but they had nothing that could compare with "Dixie."

The next morning, while the army was passing through the town and before we were relieved from duty, General Lee appeared. The female population nearly all went wild over him. All along the street through which he passed there was a perfect wringing of hands with these exclamations, "O what a grand man he is!" and "Don't you wish he was ours?"

That night we camped in Pennsylvania just over the line. As we marched along the next day we found most of the homes abandoned, the owners having fled to the mountains. Some in their haste to depart had not even closed their doors, leaving everything exposed. Nothing was molested by our men. I did not see the smoke going up from a single dwelling or any other building fired by our men while in the enemy's country. The citizens expected it by way of retaliation. Our army took only food for man and beast and exchanged old army mules for their large, overgrown horses. This was a mistake, as one mule is worth a dozen horses for military service.

On the night of June 30 we bivouacked on the summit of a high mountain, during which a heavy rain fell, drenching us.[18]

* * *

EXCERPTS FROM:
JAMES J. KIRKPATRICK DIARY,
DOLPH BRISCOE CENTER FOR AMERICAN HISTORY,
THE UNIVERSITY OF TEXAS AT AUSTEN.

1ST SERGEANT JAMES JOHNSON KIRKPATRICK
16TH MISSISSIPPI INFANTRY

June 6, 1863, Saturday: Company C went forward this morning as skirmishers. Halted our line on the railroad. Some of the company voluntarily went forward and engaged in sharpshooting. Towards evening, we are withdrawn, our brigade having orders to move up to the Howison House. Very warm. Rained this evening. Lay in trenches all night. The enemy that crossed are lying close to the mouth of Deep Run and don't seem to be at all hostile.

June 9, 1863, Tuesday:

Company C. on the front again. Cannonading in direction of Culpeper Court House. Our army has all left Fredericksburg, except our (Hill's) Corps. Most of the enemy, too, have gone.

June 14, 1863, Sunday:

Went to the front. Found no enemy on our side of the river. They re-crossed last night....Started at noon and marched to Chancellorsville. Went into camp before dark. Paid a visit to the

18. Also found in: Private Samuel W. Hankins, *A Simple Story of a Soldier: Life and Service in the 2nd Mississippi Infantry* (Tuscaloosa. Al, 2004), 46-48.

battleground of May 3rd. Very offensive now. Rights of sepulture badly performed.

June 17, 1863, Wednesday:

Reveille resounded through the woods at 3:30 a.m. Started a little after day. Very warm and tiresome marching....Several cases of sunstroke proved fatal, reminding one of Longfellow's "They, the young." Camped on the north side of Hazel River.

June 19, 1863, Friday:

Resumed our march towards Front Royal.... Halted and rested three or four hours near Front Royal. Started again and encountered a soaking rain.... Night overtakes us soon after crossing the North Fork [of the Shenandoah]. Filed off into an old field close by....Very muddy, so dark, so rainy. We rested ourselves either standing, sitting, or lying in the mud and water....

June 24, 1863, Wednesday:

Marched at sunrise. Crossed the Potomac at the ford below Shepherdstown. No pontoons as we had expected. The river is from two to four feet deep. The river is full of sharp rocks, which sometimes caused a fall, but such mishaps were rare. After crossing and redressing, we proceeded toward Sharpsburg. Halted and rest three or four hours near the old battlefield. In the evening, resumed our march through the much dilapidated looking town of Sharpsburg. Crossed the Antietam on a stone bridge near where the battle opened and camped within a mile of Boonsboro....

June 26, 1863, Friday:

Waked up before day by drums but felt very reluctant to have my slumbers this rudely broken. Lay still the drum again beat for us to fall in only a few moments later. No more delay now. Forthwith, I bundle up and call the company into line. Rained last night and this morning. Passed through Hagerstown and took the road to Middleburg, distant five miles. From thence to Greencastle [Pennsylvania]. Roads very muddy and the rain incessant. Saw some fine country, but very few sympathizers. Camped at noon two miles beyond Greencastle. A very large ration of whiskey was issued this evening. Nearly all the brigade more or less inebriated and boisterous...

June 27, 1863, Friday:

Marched early this morning. Passed through the small town of Darion, thence to Chambersburg, the capital of Franklin County, a place of some note and elegance. Our bands gave the citizens an abundance of rebel music. Took the pike here towards Baltimore. Proceeded about six miles further and camped. The citizens look very coldly on us. Their rights are very strictly guarded. An old lady got loudly cheered for running a soldier out of her garden. He had intended to get some vegetables. A retaliation for the many thefts committed in the South carefully guarded against, and any injury that may be done is without sanction of the officers. Even the orchards on the way side are well guarded. But soldiers are hard to keep under restraint....[19]

* * *

CONFEDERATE VETERAN MAGAZINE
APRIL, 1915

SERGEANT MAJOR C. C. CUMMINGS
17TH MISSISSIPPI INFANTRY

Gettysburg—Romance in Prison

In May. 1863, we started on our Northern tour in June. The last vision I had of our great commander, "Marse Robert," as we boys delighted to name him, was about the first week in June, as Longstreet's corps was moving for a grand review of the Army of Northern Virginia by General Lee at Culpeper Courthouse, which occurred on the 8th of June. As we of Company B. 17th Mississippi, were trudging along a dusty lane for this rendezvous, General Lee passed us with quite a cavalcade of his numerous staff, lie wore a long linen duster, which so enveloped his uniform as to make it invisible; added to this disguise was a gray heard about six weeks old. and he wore a broad-brimmed straw hat, evidently the art of his many lady admirers, all of which gave him the appearance of a plain old farmer. To complete the illusion, in the effort to pass the column in the middle of the road, the cavalcade became so entangled that the General was placed in the middle of his men, which gave him the appearance of being under arrest. But we boys all recognized him, for that figure could not be mistaken by the Army of Northern Virginia.

Bill Day. our Irish wit, could not resist the opportunity for fun. he was the life of the company and fed our drooping spirits on fun as a necessary tonic. Said Bill: "Boys, where did you get that old bushwhacker? What has he done? He looks like a good old man. There's no harm in him; turn him loose."

General Lee half turned his face in the direction of the speaker and with a kindly smile passed on to glory. His bright smile, after half a century, haunts me still. We saw him at a dim distance at the grand review, but never saw him at Gettysburg, as his headquarters were at

19. Also found in: Robert G. Evans (ed), *The 16th Mississippi Infantry: Civil War Letters and Reminiscences* (Jackson, MS: University Press of Mississippi, 2002), 166-87.

and around the Lutheran church, in the northern part of the city, while we were with Longstreet's Corps, on the extreme right.

We crossed the Potomac on the 20th of June at Williamsport, Md., and that evening bivouacked not far away. The next day we crossed Mason and Dixon's line, made by these surveyors as the dividing line between Maryland and Pennsylvania, which in time became the dividing line of the North and South. It was at a village called Middleburg, and the line dividing the States ran through the middle of the village. As sergeant major of the regiment. I was marching with the lieutenant colonel at the head of the column; and seeing an old man supporting himself with a cane, I called to him to draw a line in the middle of the street, marking off Maryland from Pennsylvania. He did so, and with a running jump I bounded over into Pennsylvania.

This was the 27th of June, and General Lee at Chambersburg issued his celebrated restraining order to his army, warning us against depredating on the enemy. It was a humane, Christian order from a magnificent Christian soldier and reads well in history. But my country in North Mississippi had for more than two years been overrun with Federal soldiery from Memphis so frequently that they reduced their depredations to a joke, saying my people were like a flock of geese, to be picked every six weeks.

We rested at Chambersburg, read the order, and remained there till Gettysburg opened, we getting in on the extreme right opposite the peach orchard the second day.

* * *

EXCERPT FROM:
WILLIAM COWPER NELSON, "WILLIAM C. NELSON TO J. H. NELSON" (JULY 1863).
CORRESPONDENCE 690.
(HTTPS://EGROVE.OLEMISS EDU/CIWAR_CORRESP/690)

PRIVATE WILLIAM C. NELSON
17TH MISSISSIPPI INFANTRY

You ask for a description of our march. I will endeavor to give you a slight outline. If you still have the map of Virginia you can trace it out pretty accurately. We left Culpepper C. H. June 16th and taking the route we came last Fall, passed through Sperryville, Washington, Flint Hill, Front Royal, crossing the Blue Ridge through Chester Gap, and fording the Shenandoah at Front Royal, we went on through Berryville leaving Winchester to our left. From Berryville we went to Shepherdstown passing through Charleston, and within six miles of Harper's Ferry, we could plainly see tents and flags of the enemy on Maryland Heights, the place we stormed and took last year. They have erected Forts on the heights now, and cut down all the trees that it would be impossible to take it now, as their artillery could sweep the whole mountain side, but we didn't care for Harper's Ferry and proceeded quietly on our way, they not daring to molest us. We crossed the Potomac at Shepherdstown, and proceeded on through Sharpsburg, Boonesboro, Hagerstown, Md, from there we went to Chambersburg passing through Greencastle, at Chambers burg. We found most of the stores shut, as it was not compulsory upon the merchants ever to sell. Some two or three stores were broken into, but with this exception, no damage was done to private property, the merchants who did sell, sold at (?) prices so they made a handsome profit allowing for the difference in Confederate or Yankee money.

Your affectionate Son,

Will. C. Nelson

* * *

EXCERPT FROM:
"ROBERT AUGUSTUS MOORE DIARY (VOLUME 1)."
UNIVERSITY OF MISSISSIPPI CIVIL WAR COLLECTION

PRIVATE ROBERT A. MOORE,
CO. C, 17TH MISSISSIPPI

Saturday, June 27th, 1863, bivouaced at Marion, Pa. ? 6 miles south of Chambersburg

Crossed the Pa. line about 12 o'clock, We find but few sympathisers & are not disappointed. Gen' Lee's order was read to us. He orders that private property be respected except when taken formally for use of the army. We passed through Hagerstown Md. A place of 6000 inhabitants. Also passed through Greencastel in Franklin Co. This is a very fine country between the mts.

Sunday, June 28th, 1863, bivouaced near Chambersburg Pa.

Have marched but 8 or 9 miles to-day. Have passed through Chambersburg, a city of 6000 or 7000 inhabitants & 22 miles north of Hagerstown. The souldiers are committing some depredations on private property. Have attended prayer-meeting.

Monday June 29th, 1863, bivouacked near Chambersburg, Pa.

Received orders this morning that we would remain here for the day. Gen. A.P. Hill's corps is camped near this place. Very strict orders against leaving camp. Have had chicken pie, molasses, butter-milk, pork, etc.

Tuesday June 30th, 1863, bivouacked near Chambersburg, Pa.

Moved our camp this morning & have marched 7 miles & are now near the Blue Mts. Many think we will soon move on Baltimore. Know not the whereabouts of the enemy.[20]

* * *

17TH MISSISSIPPI FOLDER,
GETTYSBURG NATIONAL MILITARY PARK

PRIVATE JOHN S. HENLEY,
17TH MS, COMPANY A

On The Way To Gettysburg

Late in the Spring of 1863, the 17th Mississippi (of which I was a member, and which formed part of Lee's Army of Northern Virginia) left Fredericksburg. We went over the plank road through the Wilderness, via Culpeper Court House, and Leesburg. We marched in ranks, four men abreast. We would march an hour and rest 10 minutes. When the rest period came, we would fall down on each side of the road. That would leave the road open, and anyone that was lagging behind, reading or walking, would try to catch up with his command. So General Barksdale came riding along. There was a tall tree standing near the road, and some of the boys began to look up into the tree pretending to see some one up there and saying "I reckon you will come down out of that tree, now, that the General has come." General Barksdale took a good look up the tree, and of course there was no one up that tree, so we all had a good laugh. The General got mad and said it was "insubordination."

We went on and waded the Potomac River, near Williamsburg. The water was cold. We moved right on into Maryland. We were marching along a road in that state and passed close to the road was one of the finest apple trees, with the finest apples on it that I ever saw. The house stood outside the garden, and an old man was sitting outside, with his chair leaned back. The boys commenced hollering comments, such as, "Come down out of that apple tree." The old man jumped out of his chair into the garden, and ran as fast as he could to the apple tree, and looked up into it, but could not see anybody. We all hollered and laughed. The old fellow went back, looking like a fool.

We took the road to Gettysburg and got there about nine o'clock the next day.

20. Also found in: James W. Silver, A Life For The Confederacy: The War Diary of Robert A. Moore, Pvt. CSA (Jackson, TN: McCowat-Mercer Press Inc., 1959), 152-53.

* * *

Confederate Veteran Magazine
January 1925

Private G. W. Bynum
2nd Mississippi Infantry

FORWARD AND BACK

By Col. W.A. Love, Columbus, Miss.

The following extracts are from one kept by G. W. Bynum, private, Company A, 2nd Mississippi Regiment, Davis's Brigade, Heth's Division, A. P. Hill's Corps, penciled on the march and in camp, that may be of interest to others than Mississippians who made the campaign.

"June 10, 1863. We have been lying in the entrenchments (at Fredericksburg) for three or four days. All quiet except an occasional shot from the artillery. General Lee has gone up toward Culpeper Courthouse with Longstreet's Corps.

"June 14. The Yankees this side of the river fell back last night, and we moved up near Falmouth, which is situated on the opposite side of the Rappahannock, two miles above Fredericksburg.

"June, 15. The enemy's pickets retired last night, except their videttes. Tom Arnold, Corporal Patrick, and myself went across the river to reconnoiter. The few videttes fell back when they saw us wading the river. When we appeared in the streets of Falmouth, I never saw a happier people. The old men and ladies happily met us with a cordial hand- shake, their eyes brimming with tears of joy. We went through the village to the enemy's camp on Stafford's Heights, and then returned to find Davis's Brigade on the march. To-night we are camped near the Wilderness battlefield.

"June 17. Arrived at Culpeper about ten o'clock and camped.

"June 18. March continued to-day. Very warm and dis- agreeable. Several of the boys were overheated and fell out of ranks, Brother Turner among them. [There were six Bynum brothers in this company, and, although often wounded, they all survived the war.] We are now camped on a high hill on the north side of the Rappahannock.

"June 19. Still on the march. Camped to-night within seven miles of Front Royal.

"June 20. Crossed the Blue Ridge and waded the Shenandoah River. Camped in three miles of Front Royal.

"June 21. Left the Winchester pike and passed through White Post, and are now camped near Berryville. General Longstreet's corps is here also.

"June 22. Rested to-day.

"June 23. Left camp this morning; passed through Berryville and Rappan, and now we are camped near Charleston, a place made famous by the hanging of John Brown.

"June 24. Passed through Charleston and are now in two miles of Shepherdstown on the Potomac.

"June 25. Crossed the Potomac by wading and passed through the battlefield of Sharpsburg, which was fought September 17, 1862. Much sign of the conflict is visible. The low mounds which cover the bones of those who fell, the furrowed ground, and scarred trees — all speak more plainly than words of that terrible conflict. I saw the ground over which we charged on that memorable occasion and the very spot where I was wounded. Sad, sad thoughts are recalled by again reviewing the old battle ground. To-night we are camped near Hagerstown, Md.

"June 26. To-day we crossed over into Pennsylvania. The people appear to be badly frightened on account of our presence.

"June 27. To-day one year ago we were fighting around Richmond. To-night a large portion of Lee's army is across the mountain. We are now camped at the base of Cumberland Mountain near Greenwood, Pa.

"June 28. Remained in camp cooking rations. Our army is pressing a number of horses into Confederate service.

"June 29. Marched across the mountain and camped near Cashtown.

"June 30. Remained in camp to-day. Raining."

* * *

DUBLIN PROGRESS
DUBLIN, TEXAS
AUGUST 8, 1913

C. C. CUMMINGS
17TH MISSISSIPPI INFANTRY

Gettysburg

(Mr. C.C. Cummings in the *Fort Worth Record*. Mr. Cummings has addressed the Confederate Veterans of Dublin.)

The night of June 30th we bivouacked near Cashtown, about six miles from Gettysburg. All that day on the march I had presentiment of evil to befall me in the coming battle. Just where the clash of arms was to be we knew not, but every private was as much aware that it was going to be the decisive battle of the great war as any general. We crossed over into the enemy's country seeking to end it at a blow and Gettysburg went down in history as the field, the fatal Flodden field. All the day of the 30th I was obsessed with this presentiment of evil and I confided my forecast to Lieutenant Colonel Fizer, saying I preferred a wound in the arm that I might escape being shot again after being down, which frequently happened when wounded in the locomotive powers. Fizer was a dressy comrade and preferred a wound in the leg, as he did not wish to mar his shape. We both got our preference on the second day in the peach orchard where we went up against Sickles' corps in that valley of death.

* * *

EXCERPT FROM:
THE VETERAN'S STORY, BY ADA CHRISTINE LIGHTSEY (ED.), 1899, 34-36.

PRIVATE R. J. LIGHTSEY
16TH MISSISSIPPI INFANTRY

After passing- through Hagerstown, Md., we took the dirt road to Chambersburg-, Pa., passing several small towns on the way. I don't remember their names; one, however, was called Germantown. The inhabitants were engaged in truck farming, their principal crop being- onions, which were very tempting- to a hungry Rebel, but we had positive orders from Gen. Lee not to disturb anything. There was one old "Reb" who could not stand to see so many fine onions without trying- some of them. So he lifted a picket from the fence, crawled in, and stooped down to pull up a very large onion. About this time an old Dutch lady whacked him on the head with a long" handled broom; he beat a hasty retreat with the old lady- close upon his heels, whacking- him every step. Finally he reached the place where he went in. As he went to crawl out she hit him with the broom handle but he held on to the onion and brought it out with him. Reaching Chambersburg, we marched through the principal street, our band playing- "Dixie." There were a few small Confederate flags displaced. They were thrown out from windows and pulled back in a hurry. When our band played "Yankee Doodle" the Stars and Stripes could be seen floating- from nearly every window in town. We moved out a mile from town and camped till the first day of July.

* * *

CLARION-LEDGER
APRIL 16, 1909

PRIVATE FRANK H. FOOTE
48TH MISSISSIPPI INFANTRY
(FORMERLY 2ND MISSISSIPPI INFANTRY BATTALION)

Our command – Brigadier General Carnot Posey's Mississippians – in June, 1863 marched with Lee's army toward Pennsylvania, meeting passing through Charleston, Va., the ladies there, in their patriotic fervor surrounded General Posey, cheered for him to dismount and by turns he was "kissed for the Mississippi Brigade." We looked on and yelled our approbation upon being kissed by proxy.

Sharpsburg, Md., was reached, and a rest given us. While sitting on the porch of a cottage with a comrade a door behind us was partially opened and a lady quietly asked us "If we do not with for something to eat?" Assenting of course, we were told to go through the gate to the rear of the home, and try not to attract too much attention. We were handed out a generous slice of bread butter and a glass of milk. A delicious lunch it was. Thanking her for the kindness we started to leave for our command when she stopped us and in a quiet dignified manner said "I have given freely to you men such as I have to give, but I assure you it is not for any sympathy for your cause to it is one I abhor and detest, but because I saw you were tired and hungry. I hope to see the Union reunited and pray it may be without any more battling for I have dear ones in the Union army fighting for its preservation." Her quiet manner and patriotism and above all her womanly feeling made a lasting impression upon me.

General J. B. Gordon in his book relates almost this identical incident as occurring to him on the Gettysburg campaign though he and staff were sitting with a family at lunch.

Passing Berryville we noticed a fat black mammy leaning on a fence and gazing intently at me as we passed by. Presently she queried us: 'is dere any more sojers behind youns?' Replying in the affirmative she broke in. 'Fore God I nebber seed so many people before, dey have his pass'n here now three days and dere is more a comin.' 'Jess see, al' dem are men and dy's all got mammies to.' The world was big to her.

Near Gettysburg I was talking with an aged lady who was with her two granddaughters, was busy seeding cherries. She asked me "where I was from," and I replied Mississippi. "Any men in your command from

Vicksburg?" I told her we had three companies in my regiment from Vicksburg. "Have you a man there by the name of Grammer?" I told her yes. "When you go back to your command, won't you please tell him to come and see me. I am his grandmother and these are his cousins. Grammer was told of it and getting permission he visited her and returned loaded with many comforts. Blood is thicker than water.

Passing through a little hamlet in Pennsylvania Greencastle the whole populace remaining there were out to see 'Lee's men' marching by. Some gazed sullenly at us, others flaunted Union flags and jibed us as rebels. Two young girls were particularly zealous in waving their colors at us., crossing them over their breast and marching with us. A rough uncouth old Johnny Reb happening alone in the rear caught the brunt of these girls demonstration and tired of it and vented his ire by saying, "Look here girls you uns had better take off these durned flags, for we rebs are hell on breastworks." A good natured laugh from soldiers and citizens greeted the girls as they retired discomfited. I published this incident in 1882, and strange to relate I shortly afterwards got a letter and photograph from Philadelphia in which the writer and his wife was then one of these girls and asked me if I could recognize her from the picture.

* * *

Clarion-Ledger
August 30, 1981

Private N. Newton Nash
13th Mississippi Infantry

Seldom did he (Nash) mention any thoughts of dying. But on June 28, 1863 – just four days before his being killed – he wrote the last letter, datelined Chambersburg, Pa. He began by telling Mollie how beautiful the landscape was in Pennsylvania and how high were the sprits of the Confederates who were invading the enemy's country.

06 28 1863 [Newton Nash Letters, 13th MS Co. I] MDAH

Chambersburg Penn, June 28th 1863

Dearest One:

You will see by the caption that we are in the enemies country; and so far has amply paid for the invasion. We are all living on the fat of the land.

Many of the people will take Confederate money ... They say we treat them much better than they expected; acknowledge that their men have acted wrong in making war upon unarmed people. There are a good many who say they are peace democrats or copperheads, have always been opposed to the war.

Private N. Newton Nash

(H. Grady Howell, Jr.)

We are about two hundred and thirty miles from Richmond, Va. So you see if we should be defeated here and have to make a retreat, we would have a hard time of it. But I will trust that to God, General Lee, and the pluck of our men.

We are between the Blue Ridge, and Allegheny mountains, the richest and most beautiful country I ever beheld. I hate to see such a

Mollie Nash

(H. Grady Howell, Jr.)

fine country in the possession of such people. We have made this long march without the loss of many men; and it is not over yet by far; for the army is still moving on. All are in the finest spirits. I think we will have the Capitol [Harrisburg] before long ...

Darling, if I should get wounded here and left behind, I have friends here that will acquaint you of the fact; but now if such should befall me; for my sake don't be uneasy about me; if not for your own, though I don't fear any thing of the kind; for I have been in many hot places; and have come out safe. My trust is still in the same God that has always shielded me from harm.

Capt. is sick with diarrhea, but has kept up by riding in the ambulance. All the rest are well. The mail boy is waiting.

Give my love to all, and may the good Lord bring this unholy war to a speedy end; and may we all be permitted to return to loved ones at home is the daily prayer of your ever devoted.

Newton

P.S. I am in fine health

* * *

VICKSBURG EVENING POST
VICKSBURG, MS
1896

SGT. CHARLES E. HUTCHINSON,
CO. H, 48TH MISSISSIPPI

Soldier Diary

June 24 – Started early and crossed the Potomac at 10 o'clock. Stopped to rest in sight of Sharpsburg. We are now fairly started into Maryland. Hope we will be more fortunate than before. Passed through Sharpsburg. Houses are closed. No Southern feeling here. At Keidersville Ditch, through which we passed, a son of General Wright was taken prisoner just ahead of our troops today. We are camped near Boonsboro.

June 25 – Started very early, passed through Boonsboro – All Union. We captured several spies in town yesterday. We are marching through a fine country and are within ten miles of Pennsylvania. Passed through Franktown; a good many Southerners in it. We are camped in sight of Hagerstown.

June 26 – Passed through Hagerstown. All Union. Did not see but one Southerner – but she gave us a wave. Middletown, through which we passed, is on the Pennsylvania line. Two or three small specimens of Confederate flags were exhibited – A very pretty town. All rabid Union. Some very pretty girls – but they looked very black at us. Marched twelve miles and camped at 12 o'clock. Raining very hard all day. Miserable souls since we left the Pike in Maryland. There has been some "pressing." It looks hard but the Army must live. I hope there will be no ladies insulted. Expect a fight in a day or two.

June 27 – Wrote home. Left camp about 8 o'clock. Marching in rear of division. Passed through Marion, a small village. Saw General Longstreet today for the first time on march. He rode into Chambersburg at the head of our brigade. The "Burg" is a very nice place of about ten thousand inhabitants. All Yankees and full of young men. Some of our boys haven't acted right, are in hopes it will be stopped. Strict orders have been issued in regard to it. We are camped near the mountains.

June 28 – Remained in camp all day. We had preaching in camp, two young ladies were present. Foraging party sent out this morning, and returned with chickens, eggs, honey, butter and onions, for which they paid in Confederate money.

June 29 – Have orders to be ready to move at 7:30 o'clock. Orders countermanded. Good news from the brave old state of Mississippi. Grant has been repulsed again with great loss; reported at ten thousand. Hurrah! for Mississippi. Have had some rain today. Expect we will move in the morning.

June 30 – Still in camp. Have had some rain today. Have heard we are the "Reserve Division." Some of our battalion came in today, that had been left behind sick. They report Yankees at Hagerstown.

CHAPTER 2

Gettysburg Day 1

Wednesday July 1, 1863

The Mississippi brigade of Brig. Gen. Joseph R. Davis was one of the first two Confederate brigades to engage the Union Army of the Potomac at Gettysburg on the sultry morning of July 1, 1863 thereby initiating the greatest three-day battle of the American Civil War. With A.P. Hill's approval, Major General Henry Heth was determined to enter Gettysburg with his division early that morning to seek out supplies and scatter any Union force of militia or cavalry present. Accordingly at 9 a.m. he ordered Davis to advance due east on the left side (i.e. immediately north) of the Chambersburg Pike while Archer's brigade did the same on the right side (immediately south) of the turnpike supported by seventeen cannons from Major William Pegram's artillery battalion.

Unfortunately Davis did not advance at full brigade strength of four regiments and, more importantly, not with his most experienced soldiers. Indeed Davis' inexperience as a brigadier general became apparent from the outset reflected in several poor decisions he made that day. With a fight expected, Davis chose to leave his second largest and most battle-hardened regiment—11th Mississippi—behind to guard the division trains in the vicinity of Cashtown (much to their chagrin). This dubious choice is perhaps even more perplexing since the 2nd Mississippi had been on detached vidette duty throughout the previous night and therefore could not be expected to be at peak fighting efficiency. The inexperienced 42nd Mississippi and 55th NC clearly represented a superior choice to leave behind. Lastly this day represented the first time Davis' brigade fought together as a unit.

Dr. Legrand J. Wilson, surgeon of the 2nd Mississippi, participated in the advance and left this description:

> **We had been ascending a gradual incline for a mile or more. When we reached the top of the hill [Wislers (Knoxlyn) Ridge] a beautiful expanse of country lay**

Dr. Legrand J. Wilson, surgeon of the 2nd Mississippi Infantry

(Mississippi Department of Archives and History)

> **before us for about a mile and a half, covered with a luxuriant growth of wheat nearly ready for the husbandman's sickle. A little to the right of a direct line was Gettysburg, a beautiful town, and beyond it a high range of hills. This is about what we took in at a glance, but we had little time to admire the beautiful landscape, dotted over with handsome farm houses, and the finest barns I ever saw. There was a puff of smoke a mile away, followed by a loud report, and a shell went hurtling over our heads. Another and another followed in quick succession. The brigade was rapidly deployed into line and advanced across the wheat field toward the battery.[1]**

This incoming artillery fire emanated from Lt. John Calef's Battery A, 2nd U.S. Artillery posted across the Chambersburg Pike along McPherson's Ridge in support of Brig. Gen. John Buford's dismounted cavalry division. Crossing over Marsh Creek this Confederate column then encountered Buford's skirmishers armed with new six-shot Spencer repeating rifles. Despite the unexpectedly heavy rate of fire from this Union skirmish line, and still raked by constant artillery fire, these determined Mississippians pressed steadily onward until reaching Herr Ridge.

Under specific orders to "take position on the left of the turnpike and, with the right resting on it, press forward towards the town," about 10:30 a.m. Davis deployed his forces with the 42nd Mississippi commanded by Col. H.R. Miller on the right, adjacent to the pike, the 2nd Mississippi in the middle under Col. J.M. Stone, and the 55th NC commanded by Col. J.K. Connally on the left flank. Davis quickly threw out skirmishers and the brigade proudly advanced in line of battle over Willoughby Run toward McPherson's Ridge, then defended by two brigades of Buford's cavalry division under Colonels William Gamble and Thomas C. Devin, fighting dismounted.[2]

Buford's stubborn resistance held Davis at bay long enough for Union infantry from the First Corps to arrive and deploy under direction of Major General John Reynolds. Running across the open fields at the "Double Quick" and loading their rifles "on the run," Brigadier General Lysander Cutler's Brigade of Brigadier General William Wadsworth's First Division was the earliest to arrive on the field and deploy — three regiments north of the Chambersburg Pike

1. Wilson, *The Confederate Soldier*, 116. The general battle narrative is derived from the following sources; Gottfried, *Brigades of Gettysburg*, 48-53; 613-18; Philip Laino, *Gettysburg Campaign Atlas*, 3rd Edition (Gettysburg Publishing, 2014) 92-104; *Rufus R. Dawes, An Officer in the Iron Brigade*, copyright 1890 (Bolton, ON, 2017) 144-53.
2. *OR* 27/2, 648-49.

Colonel John M. Stone of the 2nd Mississippi Infantry

(Public Men of Today)

(56th Pennsylvania, 76th New York and 147th New York) and two regiments south of it (95th New York and 84th New York), supported by the six cannons of Captain Hall's Battery B, 2nd Maine Artillery, which relieved Calef's battery now departing the field.

As the 42nd Mississippi and 2nd Mississippi advanced directly eastward through a tall wheatfield, the 55th NC swung north then eastward through a series of three ridges to envelope the right flank of Cutler's line near Shead's Woods. The 55th NC poured a devastating volley into the Union right flank—their first shots fired in combat — while the 2nd Mississippi under command of Col. John M. Stone fired repeated volleys from directly in front. Scores of men from the 56th Pennsylvania and 76th New York fell immediately. Colonel Connally ordered the 55th NC to fix bayonets and right wheel to gain the Union rear. During this charge the regiment's color-bearer was shot down and, caught in the moment, Colonel Connally instinctively seized the flag and rushed forward waving it aloft. Instantly he suffered serious bullet wounds in the hip and arm, the latter later resulted in amputation of the limb. When asked by Major Alfred Belo if he was seriously wounded, Connally reportedly replied — "Yes, but do not pay any attention to me; take the colors and keep ahead of the Mississippians." Elsewhere during this action Colonel Stone dismounted to climb over a tall worm fence with men of the 2nd Mississippi. While in the act, Stone was laid low by a piece of shrapnel striking his side. Private Samuel W. Hankins later wrote: "the 2nd Mississippi was fortunate in having one of the best colonels in the army, and who after the war for years made one of the best Governors the State of Mississippi ever knew." To his right Lieutenant Colonel Hillary Moseley and Major William A. Feeney of the 42nd Mississippi were both severely wounded and taken out of action.[3]

The pressure exerted by the 2nd Mississippi and 55th NC upon Cutler's two right flank regiments quickly became too severe and Wadsworth ordered them to retreat to Seminary Ridge— but not before the 76th New York lost 234 men killed/wounded/captured out of the 375 engaged and the 56th Pennsylvania lost half their strength in this deadly encounter. This left the 147th New York (another untested regiment seeing its first action) further to the west near Hall's battery

3. Charles M. Cooke, "The Fifty-Fifth Regiment," in Walter Clark ed., *Histories of the Several Regiments and Battalions from North Carolina 1861-1865* (Goldsboro, NC, 1901) Vol. 3, 297; T.P. Williams, *The Mississippi Brigade of Brig. Gen. Joseph R. Davis: A Geographical Account of Its Campaigns and a Biographical Account of Its Personalities, 1861-1865* (Dayton, OH, 1999) 86; Samuel W. Hankins, *Simple Story of a Soldier: Life and Service in the 2nd Mississippi Infantry* (Tuscaloosa, AL, 2004) 34. Stone was the longest serving Mississippi governor.

Colonel John K. Connally of the 55th North Carolina Infantry

(North Carolina Regiments by Clark)

Lt. Col. Rufus Dawes of the 6th Wisconsin Infantry

(Wisconsin Historical Society)

in an isolated and vulnerable position facing the 42nd Mississippi with the 2nd Mississippi and 55th NC on its right flank and rear. The order to retreat had been delayed due to the wounding of its commander so they fought it out. Captain J.V. Pierce in a letter to John Bachelder later wrote — "The fight was again fierce and hot; the boys on the right were falling like autumn leaves; the air was full of lead. Men fell all along the line."[4] Finally receiving the order to retreat, these bloodied New Yorkers raced for the rear hotly pursued by the onrushing Confederates. But their valiant sacrifice had enabled all but one gun of Hall's battery to escape after firing final volleys of canister.

While this action unfolded north of the Chambersburg Pike, Brigadier General Solomon Meredith's First Brigade — the famed Iron Brigade, 1829 strong, and characterized by their tall black Hardee hats pinned on one side and sporting a plumed feather — arrived in front of Herbst woods south of the pike to confront Archer's Confederate brigade. The 6th Wisconsin under command of Colonel Rufus Dawes

4. Gottfried, *Brigades of Gettysburg*, 51.

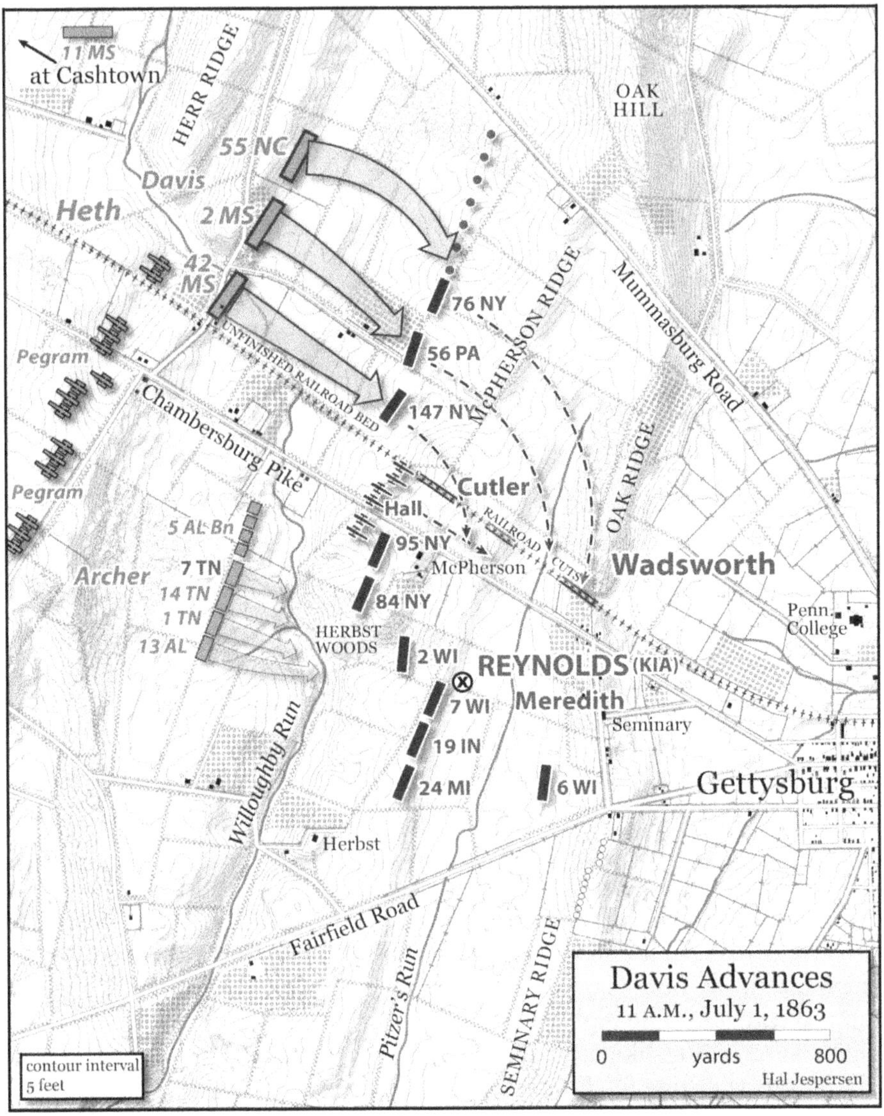

was the last regiment of the brigade to arrive and was designated as the reserve. However, Dawes was quickly ordered to proceed north at the double quick to intercept Davis' onrushing Mississippians now threatening the Iron Brigade's right flank. At the same time Cutler's two remaining regiments south of the pike changed front from north to west to repel this new threat. From the fence along the Chambersburg pike, Dawes immediately loosed a volley into the 55th NC on Davis' left flank while the 84th and 95th New York regiments fired obliquely into Davis's right flank and centre.

Captain Thomas G. Clark of the 48th Mississippi Infantry

(H. Grady Howell, Jr.)

This action now reached a critical point. Thus far Joe Davis' Mississippi Brigade, in its first battle action, had been successful in sweeping aside Buford's skirmishers and three of Cutler's infantry regiments. However, two of his three regimental commanders were out of action and his battle line was now ragged with the regiments jumbled in an undisciplined pursuit of the fleeing Federals. Davis'

Jonathan and Henry Clark of the 48th Mississippi Infantry

(H. Grady Howell, Sr.)

failure to halt and reform his ranks would now cost him dearly. At this juncture Major Belo advised Major John A. Blair, now commanding the 2nd Mississippi, to immediately charge his regiment because he realized "the side charging first would hold the field."[5] While Davis and Blair demurred, Dawes, grasping the desperate situation and under a hail of gunfire, immediately ordered the 6th Wisconsin to charge accompanied by the 95th and 84th New York regiments across 150 yards of open ground toward an unfinished railroad cut where members of the 2nd Mississippi and 42nd Mississippi had taken cover. During this action Captain Thomas G. Clark and his two young sons, Jonathan (21 yrs.) and Henry (18 yrs.), all from Company F, 42nd Mississippi were killed in the melee. As Sergeant Augustus Vairin of the 2nd Mississippi wrote in his diary — "our men thought [it] would prove a good breast work but it was too deep and in changing front the men were tangled up and confused."[6] Overcoming initial rifle fire and hand-to-hand combat here, the 6th Wisconsin quickly sealed the eastern entrance to the cut soon followed by the western exit, thereby trapping the Mississippians. Under threat of annihilation, Major Blair

5. A.H. Belo, "The Battle of Gettysburg," *Confederate Veteran*, 1900, Vol. 8, 165-68.
6. H. Grady Howell, *For Dixie Land I'll Take My Stand*, Vol. 4 (Madison, MS, 1998); Augustus Vairin, A.L.P. Diary, July 1, 1863, MDAH, Jackson.

Major Alfred H. Belo of the 55th North Carolina Infantry

(North Carolina Regiments by Clark)

personally surrendered up to 225 men and officers to Colonel Dawes. During this melee Corporal Francis A. Waller of the 6th Wisconsin seized the battle flag of the 2nd Mississippi after a tussle with color bearer Corporal Joseph B. Murphy and was later awarded the Congressional Medal of Honor for this feat.

After witnessing the debacle at the railroad cut, General Davis ordered his brigade to retreat back to Herr Ridge with the 55th NC covering the rear in the face of continued firing from the 6th Wisconsin, 84th New York, and 95th New York. There they rested during the remainder of the afternoon until assisting in the Confederate advance later in the day. At sundown the brigade was situated at the edge of town with the 55th NC positioned behind the railroad embankment in front of the Seminary. Unfortunately this regiment lost their remaining two field officers late in the day after the heaviest fighting had subsided. While waiting in reserve Lieutenant Colonel Maurice T. Smith was mortally wounded while

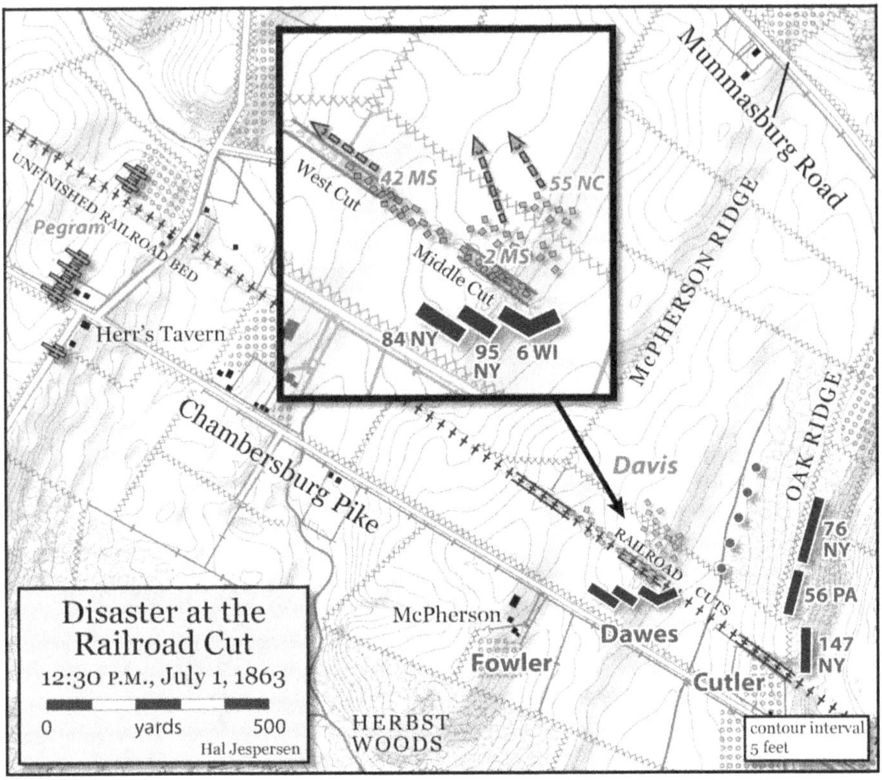

reconnoitering to the right and died later that night. Just as the combat was ending Major Belo suffered a serious leg wound which incapacitated him until 1864.[7]

While the three regiments of Davis' brigade deployed that day had inflicted horrendous casualties on three regiments of Cutler's brigade north of the Chambersburg pike, and to lesser degree on the two regiments south of it, as well as numerous casualties upon the 6th Wisconsin of the Iron Brigade, it had come at great cost. Only two of the Mississippi brigade's nine field officers escaped unscathed and remained on duty with their troops. While all three regiments had been decimated, the 2nd Mississippi fared by far the worst with only 60 men out of the original 492 present that evening. Although brigade losses were not compiled separately for this day, some estimate Davis' Brigade lost two-thirds of its initial strength.[8]

7. Charles M. Cooke, "The Fifty-Fifth Regiment," 298.
8. *OR* 27/2, 649-50; William A. Love, "Mississippi At Gettysburg," Publications of The Mississippi Historical Society, Vol. 9, 1906, 29.

In his insightful book, *The Mississippi Brigade of Brig. Gen. Joseph R. Davis*, Author T. P. Williams summarized this day's action and Joseph R. Davis' performance succinctly with these words:

> **The men under Davis and Archer had fought bravely, even heroically, yet their efforts were frustrated for several reasons. One, perhaps, was their own lack of experience. Another was the uneven generalship of Heth, who committed only half of his division, even after the dimension of the conflict was evident. Then there was Davis. For him it had been a morning of great success and greater disappointment, yet mostly it had been an education. In the desperate situation created by Dawes' unexpected onslaught, Davis lost control of a portion of his command at the unfinished railroad. It had happened to more experienced generals and, like others before him, he must learn from these mistakes. Victory in the contest deservedly went to the Iron brigade and the New Yorkers.**[9]

* * *

CONFEDERATE VETERAN MAGAZINE
MARCH 1913

CORPORAL SAMUEL W. HANKINS
2ND MISSISSIPPI INFANTRY

The following morning we marched down from the mountain to the pike that led to Gettysburg, and then on in that direction.

Between nine and ten o'clock on July 1 we halted at the foot of a hill and fronted, when our colonel (Stone) came down the line, stopping in front of each company and giving instructions. On reaching ours he remarked: "Men, clean out your guns, load and be ready. We are going to have it!" Our first lieutenant, Whitely, had been under arrest a few days for disobeying some petty order and had kept along with the company at will. To him the colonel remarked: "Lieutenant Whitely, you can take command. "Thank you, Colonel," said Whitely, obeying. We then marched to the top of the hill.

Up to this time not a report of a gun of any kind had we heard that evening.

9. T.P. Williams, *The Mississippi Brigade of Brig. Gen. Joseph R. Davis*, 91.

Samuel Hankins of the 2nd Mississippi Infantry

(Confederate Veteran Magazine)

Other regiments filed to the left until Heth's Division was all in line except the 11th Mississippi, which was on detached duty. In front of the entire line there was a large opening descending into a small valley, and across this up a slope to the edge of a line of timber. The entire opening appeared to be covered with wheat ready for the harvest. To our left we could see a long way down the valley, though to our right only a short distance could be seen on account of hills. Twenty or thirty paces in front of us Bradford's Mississippi Battery had unlimbered and been placed in line. Just as we came to a front I noticed one of the gunners directly in front of me lean backward (as did all the rest) and jerk the wire that fired the first shot as a signal for the opening of the greatest battle ever fought on the American Continent. I saw the shell when it exploded in the air over the enemy's line, which extended just in front of the skirt of timber as far as we could see both to the right and left. They were in the wheat, however, lying down, though plainly seen, while their officers rode up and down their lines. As soon as the shell from our gun exploded I noticed the smoke rise from our guns. Then came the report and shell which exploded over our heads, but did no damage.

In the meantime we had thrown out a skirmish line. J. P. Gambrell now a noted Baptist divine in Texas, was then a gallant young lieutenant and a member of our regiment, and no braver man ever lived. He was given command of our skirmishers, who soon became engaged all along the line, while the artillery duel began. We moved forward while our line of skirmishers kept pressing back that of the enemy.

After we had advanced through the wheat, across the valley, and up the slope to within a good shot of their line, they jumped to their feet and opened fire on us. We continued advancing and firing. They began to fall back until we had reached a shallow railroad cut, where they met with reinforcements. In this cut a hand-to-hand fight took place. Soon they gave way and were driven through the streets of the town. Lieutenant General Reynolds, in command of the Federal forces, was killed in front of our regiment near the railroad.

I was wounded at the railroad cut. A Minie ball struck me on my instep and broke the bones of my foot, lodging against the heel leader. I am thankful that it was not in the heel. The wound was most painful. I was taken back a few yards to where I found four others of my company wounded about the same time I was. They were Weems, Humphreys, Wilson, and Keeys. Wilson died a few minutes later. Keeys, being shot in the arm, was able to walk and passed on to the rear, leaving Weems, Humphreys, and myself. Our place of "safety" was very much exposed, as shot and shell were tearing up the earth all about us.

Now, I have given an exact statement as to the manner of which the opening of the battle actually occurred. Some historians ignore Heth's Division, which actually opened the battle. In looking over several school histories I find things which I know from personal knowledge are untrue. It seems impossible to get an impartial history of the War of the States. Gen. J. R. Davis brigade of Mississippians (of Heth's Division) including the 55th North Carolina, to which I belonged, fought every day, while at least two-thirds of the army fought only on the second and third days. When General Lee saw the cut up and worn-out conditions of our men on the third day in the field, he remarked that they should not be there.

My company entered the battle on the first day with forty-six men. On the night of the third day, after all was over, only two remained for duty. They were Rufus Jones and Berry Scott.[10]

[Ed. Note: Davis' Mississippi Brigade was not engaged on July 2nd. Hankins was separated from them due to his wound.]

* * *

The Bachelder Papers: Gettysburg in Their Own Words, Volume III, (Dayton, OH, 1995) 1699-1700.

Colonel John M. Stone
Commanding Officer 2nd Mississippi Regiment

The Second Mississippi Regiment

Tupelo Journal, Tupelo, Mississippi

The following letter addressed to Governor Stone, and his note to Col. Blair appended, explain themselves. We publish the letter in order that the survivors of that gallant regiment of Lee's gallant army - the Second Mississippi - may know that their undaunted courage, as displayed on twenty battlefields, elicited even the admiration of their foes. The facts in regard to the surrender of that remnant of the Second in that railroad cut, as will be recalled to the memories of the survivors, are as follows. The gallant Lieutenant-Colonel Dave Humphries, who was killed on the last day of the fight, had been placed in command of a large detail, guarding an immense wagon train. The chivalrous Col. J. M. Stone had been disabled by a shell early in the action on the first day. This left Major, afterwards Lieutenant-Colonel Blair, in command of the regiment. The fight was stubborn in the extreme. The brave men of the Second Mississippi had overcome the first line of the enemy had left a line of its dead where it fought; the Second then came in contact with the troops mention in Col. Dawes' letter. The railroad cut was deep and long, and right in front of the Confederate line of battle. When the gallant regiment with other Confederates got into the railroad cut, the charge alluded to in the letter was executed, and the Federal lines drawn around and in the rear of our troops. Under these circumstances the surrender was made. The letter proves Col. Dawes a chivalric officer and high-toned gentlemen. He will find the gallant Major to whom he alludes - his foe in 1863 - his friend in 1878 - and as willing as he to say in regard to a true Union Esto perpetua.

10. Also found in: Samuel W. Hankins, *A Simple Story of a Soldier: Life and Service in the 2nd Mississippi Infantry* (Tuscaloosa, Al, 2004).

LETTER OF COL. JOHN MARSHALL STONE
STATE OF MISSISSIPPI
EXECUTIVE DEPARTMENT
JACKSON, MISS. MARCH 27, 1890

COL. JNO B. BACHELDER
WASHINGTON D. C.

Sir:

Replying to your letter on the 15th ult. Enclosing a copy of H. R. 4972 I have the honor to say that my recollections of the formation and movements of Davis' brigade on the morning of the 1st of July at Gettysburg are so vague and unsatisfactory that they can be of little value. According to my recollections Davis brigade formed on the high ground north west of the city, overlooking the undulating fields between the point and the city of Gettysburg, with a body of timbered land to the left and nearer Gettysburg. The brigade advanced with my regiment (2nd Miss.) on the right. The other regiments (42nd Miss., 55th N. C.) were to my left but I do not remember the order. The 11th Mississippi Regiment was not with the brigade on the 1st. We advanced through open fields and, according to my recollection, in a south easterly direction first, and then almost south. I was riding along a lane, and had just crossed a branch in rear of my regiment when its right became engaged. Only the fence (post & rail) I left my horse to go on foot to right of my regiment and was shot off the fence as I was climbing over. The whole regiment was soon engaged, many killed and wounded, and a large number reached the Railroad Cut, who with the Major of regiment, Jno. A. Blair, were made prisoners.

I may be in error as to the position of my regiment, but think it was on the right. I confess, that when in battle, I known but little beyond my immediate command and that which lay immediately in front. I am therefore, a poor historian, I could not give the distance from the point at which my regiment formed to the town, but think it was 1 ½ to 2 miles.

Major A[lfred] H. Belo, Editor Galveston News, Galveston Texas, was Major of the 55 N. C. Regiment, and I think was in the battle and if so can give you much information. Col. [John K.] Connaly, commanding that regiment, lost an arm there, but I cannot give you his present address.

Gen. Joseph R. Davis who commanded the brigade resides at Biloxi Miss. Col. Jno. A. Blair, then Major 2d Miss. Regiment is at Tupelo Miss.

All the troops of Davis brigade were in the battle of the 3d day, and went in immediately on the left of Pickets [sic] Virginia Division, though I was not with them.

Respectfully

Your obt servant

J. M. Stone

* * *

CLARION-LEDGER
JACKSON, MS
JUNE 14, 1876.

1ST SERGEANT AUGUSTUS L.P. VAIRIN,
CO. B, SECOND MISSISSIPPI
JULY 1, 1863

We saw a Yankey division which had captured Pettigrew's [Ed. Note: Archer's] Brigade on our right advancing on an old Railroad cut on our right. Our men thought [It] would provide a good breast work but it was way too deep and in changing front the men were tangles up and confused. There we lost a number of men? I did not go into the cut seeing its danger and continued all I could to get out by the right flank but those on the left were surrendered by Major Blair. Those of us who got out of the difficulty fell back ? others of our troops came up in time and we gained that day's battle ? but our regiment was reduced fearfully.

Augustus Vairin, A.L.P. Diary,

July 1, 1863, Mississippi Dept. Archives and History, Jackson

Colonel Rufus R. Dawes

6th Wisconsin Infantry

Marietta, Ohio, June 1, 1876.

To Hon. J. M. Stone:

Dear Sir: I have a suspicion that you were Colonel of the Second Mississippi Confederate Volunteers. If I am in error, I am sure I can give no offense, for that noble body of soldiers could have been led by none other than a brave man. At the battle of July 1st, near Gettysburg, I commanded the 6th Wisconsin Volunteers. We charged upon the Second Mississippi, posted in a railroad cut. In that charge forty of our men were killed and one hundred and sixty wounded. You see I have reasons to remember and respect the Second Mississippi. At

Sergeant Augustus L. P. Vairin of the 2nd Mississippi Infantry

[Photo enhanced by Forever Studios. The hairline on the original was greatly faded, so the appearance of Vairin's hair is based on the software's and enhancement artist's best understanding.]

(History of Tippah County Mississippi)

the railroad cut I received from Major Blair the surrender of seven officers and two hundred and twenty-seven men. As the record shows, each of these men had shot his man before they yielded. They have no cause to blush or we to exult. But that bloody struggle is to us a common memory, and it would give me great measure to know where I might find some of the men who stood against us, that we might compare notes and recollections. Now if you could tell me where I could communicate with Major Blair, or any other officer, I would be greatly obliged.

We fought the Second Mississippi in the corn field in front of the Dunkard Church at Antietam. There they drove us, and we barely saved by a hand a battery of six twelve pound howitzers, planted in front of some hay stacks. You will remember this place well, if you are Colonel Stone of that regiment.

Very respectfully,

(Signed) R. R. Dawes.

Dear Colonel: The foregoing is a copy of a letter I have just received. Thinking it might recall to you some interesting events. I make and send you a copy. I have answered the Colonel that he can communicate with you at Tupelo.

Your friend,

J. M Stone.

* * *

EXCERPT FROM:
RUFUS ROBINSON DAWES, *SERVICE WITH THE SIXTH WISCONSIN VOLUNTEERS* (MARIETTA, OH: E.R. ALDERMAN & SONS, 1890), 159-161.

COLONEL RUFUS R. DAWES, 6TH WI INFANTRY
LETTER TO MARY BEMAN GATES
JULY 4TH, 1863, 12 M

I am entirely safe through the first three of these terrible days of this bloody struggle. The fighting has been the most desperate I ever saw. On July 1st, our corps was thrown in front, unsupported and almost annihilated. My regiment was detached from the brigade and we charged upon and captured the second Mississippi rebel regiment. Their battle flag is now at General Meade's headquarters, inscribed as follows" "Captured by the sixth Wisconsin, together with the entire regiment, kept by Sergeant Evans for two days, while a prisoner in the hands of the enemy.

The Sixth has lost so far one hundred and sixty men. Since the first day we have lost only six. O, Mary it is sad to look now at our shattered band of devoted men. Only four field officers in the brigade have escaped and I am one of them....

Following is an excerpt from Colonel Dawes' after-action report:

Headquarters Sixth Wisconsin Volunteers, Line of Battle before Gettysburg

July 4th, 1863

Sir: I have the honor to report that the accompanying battle flag of the second regiment of Mississippi Confederate Volunteers was captured by the regiment under my command under the following circumstances; shortly after the opening of the action on the morning of July first, my regiment by command of General Doubleday, detached from the brigade and ordered to support the right of the division (Wadsworth's) which was being forced back and outflanked by the enemy. I moved as rapidly as possible upon the advancing line of the enemy joining with the fourteenth Brooklyn and ninety-fifth New York on my left. A brisk fire was opened throughout our line which soon checked the enemy, and forced him to take refuge in a railroad cut. I ordered a charge upon the cut. The men moved forward, well closed and on a run. When our line reached the edge of the cut, the rebels ceased firing, and threw down their arms. Major Blair commanding the regiment in my front, the second Mississippi, surrendered his sword and regiment. The battle flag was taken before the surrender by Corporal F. Asbury Waller of company 'I' and sent to the rear in charge of Sergeant Wm. Evans of company 'H' who was badly wounded. The sergeant was taken prisoner by the enemy and held for two days in Gettysburg. With the assistance of some ladies, whose names I have not learned, he successfully concealed the color and finally when the enemy retired, brought it safely to the regiment. I have the honor to be, very respectfully,

R.R. Dawes, Lieutenant Colonel commanding sixth Wisconsin Volunteers

* * *

Arkansas Democrat Gazette
January 5, 2014

Corporal William B. Murphy, Co. A, 2nd Mississippi
Letter dated September 28, 1905 re Gettysburg

All the details and incidents are just as fresh in my mind as if they occurred only yesterday.

I as color bearer was about 50 paces east of the cut and on the side towards Gettysburg. There was no cut at all where I was. The ditch where I crossed the railroad was not more than 2 feet deep. Our brigade stopped in the ditch, and the cut to my right, the most of my regiment on my left.

I and my color guard were about 10 paces south of the railroad. I did all in my power to prevail on the boys to come on and take the battery, not knowing at the time that we were overpowered by those regiments of the enemy in our front and left rear so we made a desperate struggle to save our colors. My color guard were all killed or wounded in less than three minutes and also my colors were shot more than a dozen times and the flagstaff was hit and splintered in two or three places.

About this time the enemy made a desperate charge on us from the Cashtown pike and our brigade poured a deadly fire into them all the time they were on the charge, but it did not seem to check them on their run. Just then a squad of soldiers made a break and rushed for me and my colors, there was 8 or 10 killed and mortally wounded before they succeeded in taking the colors from me. One lieutenant rushed upon me to take the flag and one of my guard by the name of John Guillen shot him through the breast and he fell, then a very large man rushed upon me while I was tearing the flag from the staff and trew me down and took the flag from me (I have since learned that this man's name was F.A. Waller, Co. "I", 6th Wisconsin Reg't).

When I was at the Confederate reunion in Richmond, Va, in 1896, I visited the battlefield at Gettysburg, Pa., and I was on the very spot of ground on which the deadly struggle was made around the colors, and it made me feel very sad indeed to think that I stood on this spot of ground 33 years before that in a death struggle over the colors of my regiment.

Letter to Dr. F.A. Dearborn, dated August 29, 1902

I am proud of being a rebel soldier and that I was color bearer of the 2nd Mississippi Volunteers in the late Civil War between the states. But there is no North, nor South to-day with me, and it was a great blessing for the American people that we had the war, for now we are all one and live under the best government in the world.[11]

[Ed. Note: This battle flag was displayed in Wisconsin in 1864 and then stored away by the War department in Washington, DC. In 1905 by action of Congress it was finally repatriated back to Mississippi and displayed in the Old Capitol Building in Jackson. MDAH subsequently loaned the flag to the Jefferson Davis Presidential Library at Beauvoir where it survived hurricane Katrina in 2005, having been placed in a vault. One week later it was returned to Jackson where it is now on display.]

11. Alyson Hoge, "Corporal Murphy's Flag," *Arkansas Democrat Gazette*, January 5, 2014; http://www.arkansasonline.com/news/2014/jan/05/corporal-murphys-flag-20140105 (accessed 19 April 2023).

* * *

CLARION LEDGER
JACKSON, MS
JULY 20, 1893.

PRIVATE WILLIAM B. MURPHY, CO. A, 2ND MISSISSIPPI
MOORESVILLE, TEX., JUNE 20, 1892.

Gov. John M. Stone, Jackson, Miss.:

My Dear Col. – I see in the notice in the daily papers of your call for a reunion of all the Confederate Comrades to assemble in Jackson, Miss., on the 12th of July. There is nothing that would give me more pleasure than to meet you all on that day, but it is impossible for me to be there with you, and all comrades of the old Second Mississippi Volunteers. You will remember that I am the only one that ever gave up the colors of the old Second to a Yankee, and that was on that memorable day, July 1st, in the railroad cut at Gettysburg after a bloody and hand to hand fight. They got my flag, but they paid very dearly for all the honor they received at the hands of our men, who killed some four or five of the Yankees after Col. Dawes, Six Wisconsin, and Major John A. Blair had given orders to surrender and cease firing.

Those days still hold a deep impression in my mind as if it had just occurred yesterday, though it has been twenty-nine years since the battle of Gettysburg, and on last May, 30th and 31st, thirty years since the Chickahominy Swamp, or battle of Seven Pines, was fought. Just after dark our regiment ran upon the 55th Ohio, and they formed in line in the rear of their tents and the commander gave orders to fire by the bugle, when you ordered your men to lie down. We were so near them at the time when they fired that I could almost see the whites of their eyes by their own light; and your horse was pierced with eighteen bullets, and then I was the first to come to you and help get that horse off of you, and you would not speak for fear I was a Yankee. After you were released, I went around to the tents of the enemy, (the colonel's tent) and took his boots and spurs and all the good grub and a pair of blankets, and then I brought two prisoners to you, and you ordered me to take charge of them and deliver them at headquarters.

I have received a letter from Gen. R. R. Dawes, who commanded the 16th Wisconsin at the battle of Gettysburg. Will enclose to you to be read at the reunion of the 2nd Mississippi. I send greeting to all the bloody old 2nd Mississippi Volunteers.

With best wishes, I remain,

W. B. Murphy

* * *

Vicksburg Herald
April 30, 1881.

Captain J. E. Marston
6th Wisconsin Infantry, Co. E

Here's Your Sword, Captain.

To the Editor of the Sunday Telegraph:

Appleton, April 23. – Some time since you asked for little army events from old soldiers. Anything in this line, in the shape of unwritten history, will be acceptable, I suppose. What I have to say this time will more particularly concern a captain of the Second Mississippi regiment. The Sixth Wisconsin took this regiment at the point of the bayonet in a charge at the battle of Gettysburg, on the first of July, the order for which I gave myself by the request of the color guard, which was attached to my company. This was the bloody charge in the railroad cut, in which we found the Second Mississippi. In this charge I ordered a captain to give me his sword, or hand it to Lyman White, of my company. Lyman, in the line of his duty, was about to bayonet the Captain, but as I detected a command, from this officer ordering his men to surrender, at this instant, I sprang and caught Lyman's gun and saved the Captain's life. Now, my soldier boy Lyman is no more on earth. His mother and father are also dead. I have the sword in my possession. Now, as I call to mind this manly appearing officer I have but one wish, and that is, if he be living, to return to him his sword, as I drew my sword not in hatred. If this communication should reach this officer and he will inform me of an inscription on the scabbard as it was presented and by whom, with any other little incident that he may remember in connection with this charge. I will return him this sword and at the same time, will give your paper a further account.

Mississippi papers please copy.

J. E. Marston,

Captain, Co. E, Sixth Wisconsin.

* * *

Excerpt from *Blind Man on the Warpath*
1905

Private Wiley Heflin, Co. D, 11th Mississippi
Gettysburg, July 1, 1863

The column moved forward and in a short time we began to hear the guns. We well knew that the two great armies were meeting. Right here I want to say that I was in every battle that my company was in, from the beginning of the war. Never had been wounded sufficient for a furlough. I had been making my brags to the boys that I was bomb proof and a Yankee ball had never been moulded to kill me. I don't claim that I was brave, but during the day I was miserable, hearing the guns and cannons booming. I told my Captain Prince, that if the regiment didn't move forward the next morning, that I would run away and go to the Second Mississippi: he just answered, "All right, Heflin."[12]

* * *

"Peel, William H. Company C, Diary 1863-1865"
Mississippi Department of Archives and History
Jackson, Mississippi

Lieutenant William H. Peel
11th Mississippi Infantry
July 1st

"Roll up your blankets" was the order of the morning. Companies & Regts were soon formed. Somewhat to our astonishment, the other three Regts moved off leaving us behind. The mystery was solved, after a short time, by the receipt of the rather pleasant information that we had been detached for the purpose of guarding the wagon train. We moved a few hundred yards up the road, & to a position contiguous to the wagon yard, where we were ordered to "Rest." There was a fair prospect for rain, in view of which we fixed up shelters, as best we could in an open field. The morning was unusually quiet. The occasional report of a distant cannon was almost all that disturbed its gloomy stillness.

Gloomy, in consequence of the clouds that hung around us & doubly saddened by the thoughts + recollections enkindled by the booming of the heavy guns, that ever & anon broke in upon our hearing.

12. Wiley Pierce Heflin, *Blind Man on the Warpath*, (Philadelphia, MS, 1905), 24.

Lieutenant William H. Peel of the 11th Mississippi Infantry

(Mississippi Department of Archives and History)

Troops were passing rapidly & as if with one accord - silently to the front. Gen'l Posey's Brigade passed us about noon. Being from the same State with ourselves, there was a general meeting of friends & acquaintances in the midst of which I had the pleasure of meeting my three cousins in the 19th Regt.

The cannonading seemed to increase as the day waxed older & ere the evening had far advanced, the rapidity of the reports of resounding ordinance & the frequent outbreaks of rattling musketry told us but too plainly that the dreadful work had begun in interest. Towards the middle of the afternoon the engagement became pretty general. There was a hill near where we were halted, from which we had a tolerable view of the surrounding country. Intermediate woods shut the armies out from our sight, but we could see long chains of smoke, as they curled above the trees & were able, by the use of an inferior glass, to form some idea of how the day was going on this portion of the field. We watched with anxious eyes the up-curling wreath + saw with relief, that the farther appeared to recede, while the nearer seemed steadily to advance. The relief was all the greater & mingled with a degree of soldierly pride, as we knew this smoke to emanate from our Division, commanded by Gen'l Heth.

This Division, & a couple of Brigades of that of Gen'l Pender, were all of A. P. Hill's Corps that were engaged thus far. They occupied center.

We had no visible evidence of the doings of Genl Ewell, who commanded on the left, but the thundering of his artillery, & the continuous roar of his small arms told more plainly as they seemed to advance, than words could have done, that his veterans were wreaking a terrible vengeance for their late, & much loved Jackson.

Genl Longstreet whose position was on the right had not yet reached the field with the head of his column. His Corps was at this time passing us, moving in quick & silent march.

Twilight began to spread its veil abroad, & as darkness closed in around us the carnage ceased as if by mutual consent, leaving us in possession of the field on portions of which we had driven its enemy some three miles, & that awful quiet which ever follows the tremendous exertions of the powers of Heaven, Earth or Hell settled itself down upon Adams County.

A Courier came from the front with tidings from our Brigade. It had been hotly engaged & had suffered severely in killed & wounded, while several Co's of the 2d Regt had been cut off & newly all captured. We had the consolation of knowing, however, that all grades, from the Gen'l down, had acquitted themselves creditably.

Barksdale's Brigade of Mississippians was now passing. Going out to the road I had just succeeded in finding Radford & Lundy Gunn, when the color bearer came up & told me our Co. was going on picket. Exchanging a few words of salutation encouragement, etc. & congratulating Lundy on his late election to Lieutenancy we shook hands, perhaps for the last time in this world. Hastening back I found that several Co's mine among them, had already left for their respective picket posts. Learning the direction my Co. had taken, I gave chase in the darkness & after a race of about a mile overtook it. Proceeding about half a mile farther, we halted in a lane.

Barricading the road with strong rail fences + posting the necessary Videttes, the balance of us betook ourselves to sleep. Scarcely daring even to hope for an uninterrupted repose, since we were imminently threatened with a repetition of the day's rain.[13]

* * *

13. Ellen Sheffield Wilds, Ed., *Far From Home: The Diary of Lt. William H. Peel 1863-1865*, (Bloomington, IN, 2009,) 53-75.

Private David Holt of the 16th Mississippi Infantry

(Louisiana State University)

Excerpt from:

A Mississippi Rebel in the Army of Northern Virginia: The Civil War Memoirs Of Private David Holt

Private David E. Holt

16th Mississippi Infantry

South from Chambersburg, we were marching when an order came that every man must stay in ranks, and keep files closed up. That meant business. Then too, the "grapevine" telegraph became active. We learned the position of Ewell's corps, and we knew that we were to the left of Longstreet's corps. Soon news came that Heth's Division had struck the enemy, and we heard the distant cannonade.

We were ordered to load at will. The effect was magical. Every man tightened up and was on the alert. No need then for orders to keep the ranks closed up, for every man had become a good soldier by instinct.

We came closer to the firing and, at the same time, began to come upon evidences of the struggle. We passed through the edge of the town of Gettysburg and came out in line of battle of Seminary Ridge. We took the places of skirmishers from Heth's Division and advanced to the brow of the hill and halted.

It was not a fresh movement, but a continuation that progressed favorably to us. We had the Yanks going and now the thing to do was to keep them going. Yet we halted. We looked across a valley that was about a mile wide and saw Cemetery Ridge and Yanks. They had no fortifications and we could have easily taken the position from them, as they were giving way.

The regiment ordered us to stack arms. We never stacked arms except at time of a drill and never before on the field of battle. Fortunately, I was on the skirmish line, and not ordered to stack arms. There was a near rebellion. We cussed and discussed the strange order.

"Who commands this part of the line?" someone asked.

"General Longstreet," was the answer.

"Why, he is the fellow who stayed down on the peninsula and could not be gotten up in time for Chancellorsville! Why doesn't Anderson move the division forward without awaiting orders from such a man?"

The stack broke and the men took their guns. Most of the officers believed that it was a temporary delay only. But we were put to digging trenches. The artillery [unlimbered] in front of us at the brow of Seminary Ridge and trained on Cemetery Ridge opposite.[14]

* * *

VICKSBURG EVENING POST
VICKSBURG, MS
1896

SGT. CHARLES E. HUTCHINSON, CO. H, 48TH MISSISSIPPI

Soldier Diary

July 1 – Left camp near Chambersburg early. Passed through Fayetteville and were soon climbing the mountains; and when at the top heard cannons ahead of us. Marched on, and found a fight progressing near the town of Gettysburg. A good many coming out wounded, and a large lot of prisoners. Hear that Hooker has been superseded by Meade, and that we have killed Gen. Reynolds. We have driven them some distance today. Have heard that we took six thousand prisoners.

* * *

EXCERPT FROM:
THE CONFEDERATE SOLDIER (1902) BY LEGRAND J. WILSON
LEGRAND J. WILSON
SURGEON, 2ND MISSISSIPPI INFANTRY
JULY 1, 1863

Near nightfall I was ordered to see that all the dead were buried, and all the wounded sent back to the hospital. This forced me to go all over that horrid field. The Pioneer corps soon came, and I selected a place to bury our dead, when they went to work digging the trenches. My litter corps brought in the dead as rapidly as possible. The poor, wounded Federals were crying piteously for water in every direction. We kept our canteens filled and administered comfort to as many as possible. This was my first experience on the battlefield after the fighting, and it was horrible beyond description. If every human being could have witnessed the result of the mad passions of men as I saw that night, war would cease, and there would never be another battle. "It must needs be that offenses come, but woe unto that man by whom they come."

14. Thomas D. Cockrell, and Michael B. Ballard, eds., *A Mississippi Rebel in the Army of Northern Virginia: The Civil War Memoirs Of Private David Holt*, (Baton Rouge, LA: Louisiana State Press, 1995), 194-95.

CHAPTER THREE

Gettysburg Day 2
Thursday July 2, 1863

William Barksdale's Mississippi Brigade finally arrived at Marsh Creek on the western outskirts of Gettysburg at midnight on July 1st after a tedious, stop and go march of fifteen miles, forced to follow General Ewell's baggage train from Greenwood over a dusty roadway decorated with the excrement of thousands of horses and mules. The next morning about 11 a.m. Lee finalized his battle plan for the coming day based upon a faulty reconnaissance conducted earlier that morning by his Staff Engineer, Captain Samuel Johnston. Lee incorrectly concluded that the Union left flank was "in the air," unprotected in the vicinity of the Trostle farm 1.5 miles to the south near the Emmitsburg Road. Lee planned to have Longstreet's two divisions—McLaws and Hood (Pickett was still guarding the Confederate rear at Chambersburg)— attack northward astride the Emmitsburg Road at the Sherfy farm. As this surprise attack developed, the five brigades of Major General Richard H. Anderson's Division of A.P. Hill's Third Corps would mount an *en echelon* attack eastward in support. Almost immediately there was confusion as Longstreet, McLaws, and Lee studied their map with Lee pointedly correcting Longstreet that his attack was to be directed parallel to — not perpendicular to— the Emmitsburg Road. Lee also instructed Longstreet to reach the launch point without being seen by Union signal stations on Little Round Top (which ultimately necessitated a frustrating and time-consuming countermarch).[1]

McLaws and Hood did not arrive at the Sherfy peach orchard until 4 p.m. With Kershaw's South Carolina brigade in the lead, followed by Barksdale's Mississippians, and expecting to be well beyond the Union left flank, they were totally shocked at what they saw upon emerging

1. Harry W. Pfanz, *Gettysburg: The Second Day* (Chapel Hill, NC, 1987) 106-07; Eric A, Campbell, "The Key to the Entire Situation—The Peach Orchard, July 2, 1863," The Second Day at Gettysburg, 2006 Gettysburg National Military Park Seminar, (Gettysburg, PA, 2008), 150-51; Major-General Lafayette McLaws, "Gettysburg," Southern Historical Society Papers, Vol. 7, 68.

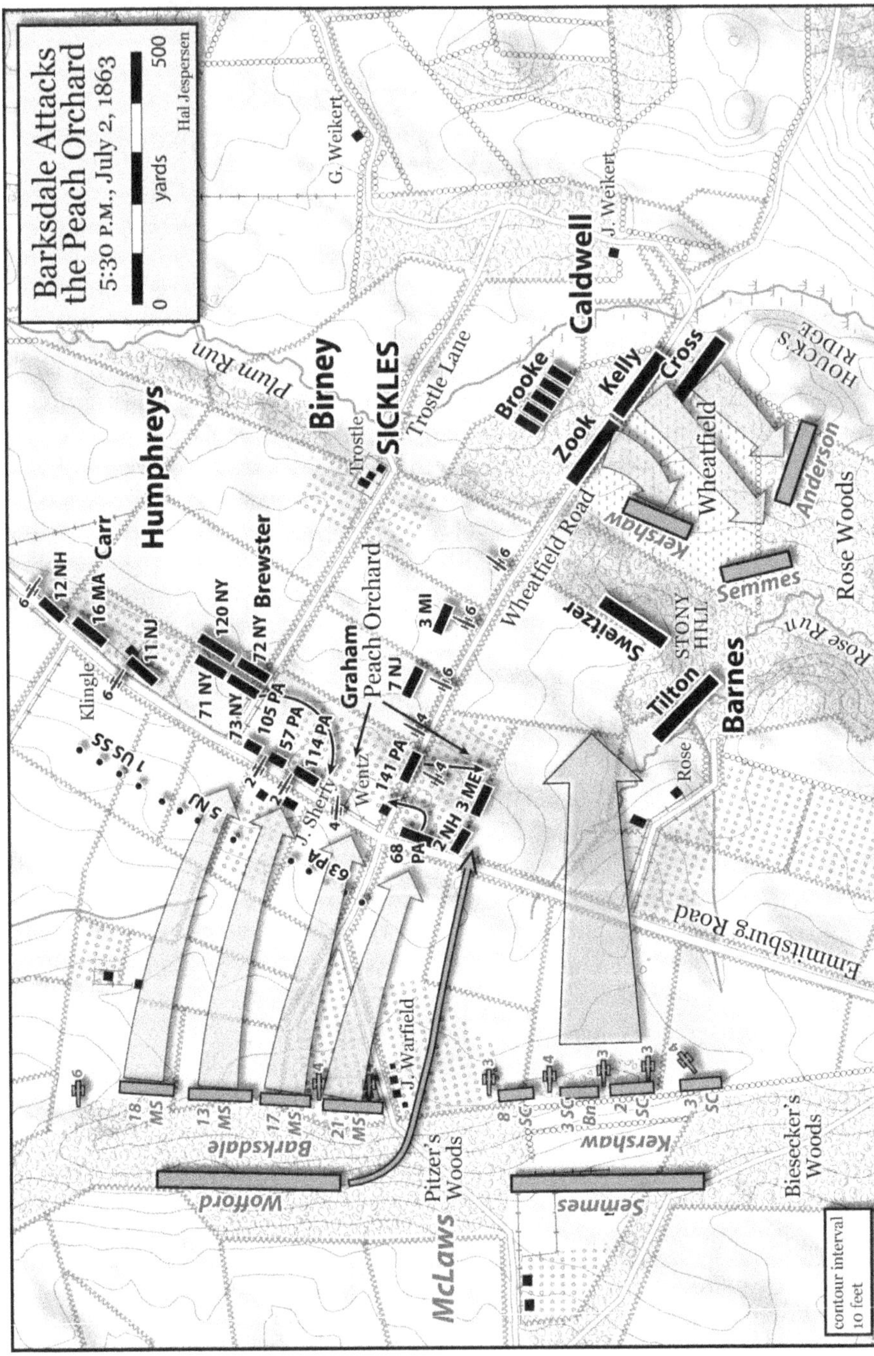
Barksdale Attacks
the Peach Orchard
5:30 P.M., July 2, 1863
0
yards
500
Hal Jespersen
G. Weikert
J. Weikert
Caldwell
Brooke
Kelly
Cross
Zook
Houck's Ridge
Wheatfield
Wheatfield Road
Anderson
Kershaw
Semmes
Rose Woods
Rose Run
Sweitzer
Stony Hill
Tilton
Barnes
Rose
Plum Run
Birney
SICKLES
Trostle
Trostle Lane
Humphreys
Carr
12 NH
16 MA
11 NJ
120 NY
72 NY
Brewster
71 NY
73 NY
105 PA
57 PA
114 PA
Graham
Peach Orchard
7 NJ
3 MI
141 PA
3 ME
2 NH
68 PA
63 PA
Wentz
J. Sherfy
Klingle
1 US SS
5 NJ
Emmitsburg Road
J. Warfield
18 MS
13 MS
17 MS
21 MS
Barksdale
Wofford
Pitzer's Woods
McLaws
8 SC
3 SC
3 SC Bn
2 SC
Kershaw
Semmes
Biesecker's Woods
contour interval
10 feet

Terrific fighting at the Peach Orchard - Illustration

(Library of Congress)

from Pitzer's Woods. McLaws later wrote: "... the view presented astonished me, as the enemy was massed in my front, and extended to my right and left as far as I could see," while Colonel Benjamin G. Humphreys of the 21st Mississippi left this description: "... it was unmistakably anything but the [enemy] flank. It was a formidable compact line of frowning artillery and bristling bayonets."[2]

During the intervening hours the military situation near the Peach Orchard plateau had drastically changed due to Major General Daniel E. Sickles' unilateral decision—without orders— to advance his III Corps about one mile westward to the high ground along the Peach Orchard plateau from his assigned position extending Major General Winfield Hancock's II Corps line along Cemetery Ridge. This maneuver resulted in a new unorthodox and vulnerable defensive line. The left flank rested on the boulder-strewn rock outcrop of Devil's Den. The line then proceeded diagonally in a northwesterly direction

2. McLaws, "Gettysburg," 70; B.G. Humphreys to McLaws, January 6, 1878, in McLaws Papers, Southern Historical Collection, UNC.

Captain Isaac Stamps of the 21st Mississippi Infantry

(Journal of Wilkinson County History)

along Houck's Ridge through the Wheatfield to the Peach Orchard. This half was defended by Major General David Birney's 1st Division. At the Peach Orchard salient the line made a 90 degree turn and ran along the east side of the Emmitsburg Road connecting with Sickles' 2nd Division under Brigadier General Andrew A. Humphreys north of the Sherfy farm buildings. Humphreys' line then extended northward along the roadway with its right flank "in the air" near the Klingle house (unconnected to Hancock's left flank well to the rear along Cemetery Ridge).

Despite standing orders to attack, which were repeated by Longstreet, McLaws' lead brigades sought shelter along Warfield Ridge (the southern extension of Seminary Ridge). McLaws later elaborated —"Kershaw, a very cool judicious and gallant gentleman, immediately turned the head of his column and marched by flank to the right, and put his men under cover of a stone wall," while "Barksdale, the fiery, impetuous Mississippian following, came into line on the left of Kershaw, his men sheltered by trees and under a gentle declivity." Seeing the dire situation for himself, Longstreet immediately rescinded the attack order and, given these new developments, desperately

Colonel Benjamin Humphreys

(Library of Congress)

attempted to modify Lee's battle plan while retaining its essential components and intent.[3]

Longstreet then ordered Hood's division to proceed southward until reaching a point beyond the actual Union left flank. Then Hood's division would attack northeastward followed by McLaws' division attacking eastward at the Rose farm (Kershaw's South Carolina Brigade backed by Semmes' Georgia Brigade) and the Peach Orchard (Barksdale's Mississippi Brigade backed by Wofford's Georgia Brigade) in an en echelon attack by brigade. Hood's extended march to the right resulted in a further delay which had deadly consequences for McLaws' Confederates. Both sides recognized that the Peach Orchard salient was the key to the impending action and dispatched additional artillery support to this sector which soon erupted into the greatest counter-battery "gunfight" of the entire war — eclipsing the "artillery hell" of Sharpsburg. Given the clear lines of sight and short range, battery targeted battery, while any unfortunate infantry units hunkered down nearby suffered horrendous collateral damage.

Unfortunately for the Mississippians, Barksdale's brigade suffered this consequence in the extreme due to Longstreet's insistence on

3. Ibid.

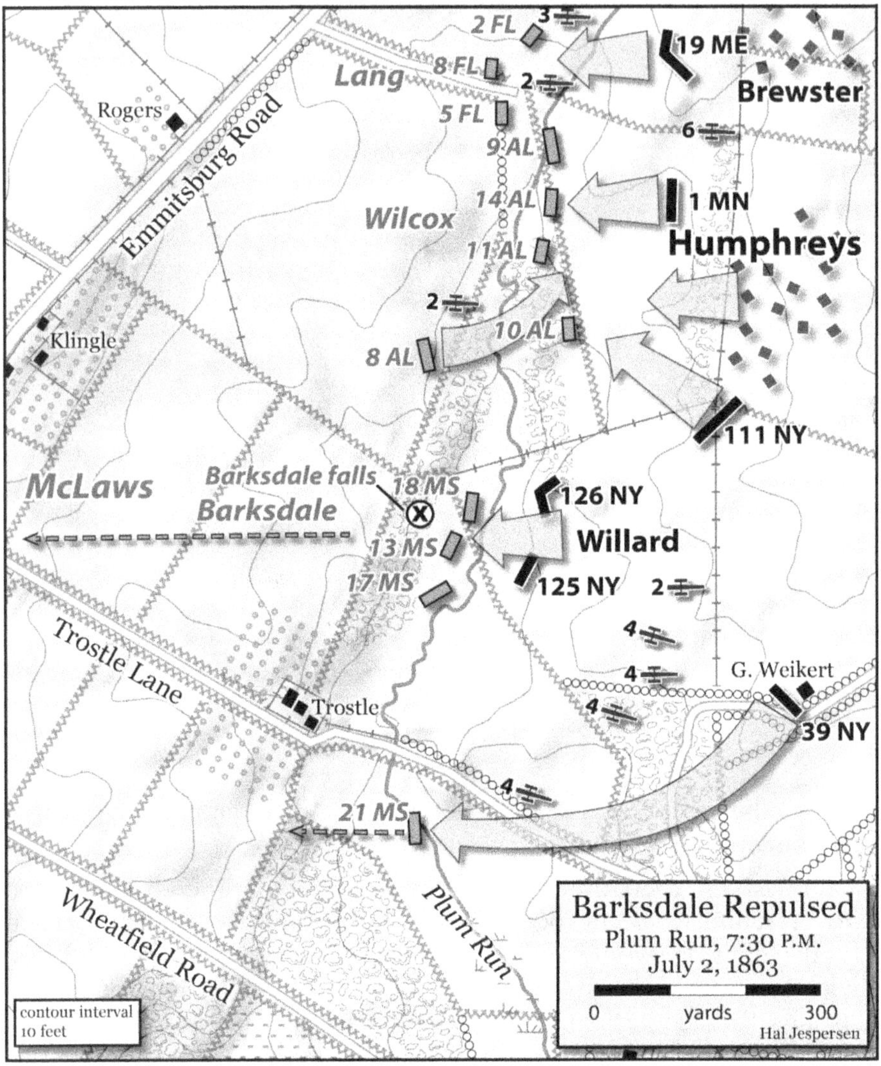

posting two Confederate batteries *directly in front of* Barksdale's position. McLaws protested to Longstreet, but to no avail: "General, if a battery is placed there it will draw the enemy's artillery [fire] right among my lines formed for the charge, and will of itself be in the way of my charge, and tend to demoralize my men." The severe pounding in front of Barksdale's position soon also took a severe toll on the Confederate batteries posted there with tragic results for several members of the 17th Mississippi. Artillery Commander Edward Porter Alexander later recalled:

> **so accurate was the enemy's fire, that two of my guns were fairly dismounted, and the loss of [artillery]men was so great that I had to ask General Barksdale, whose**

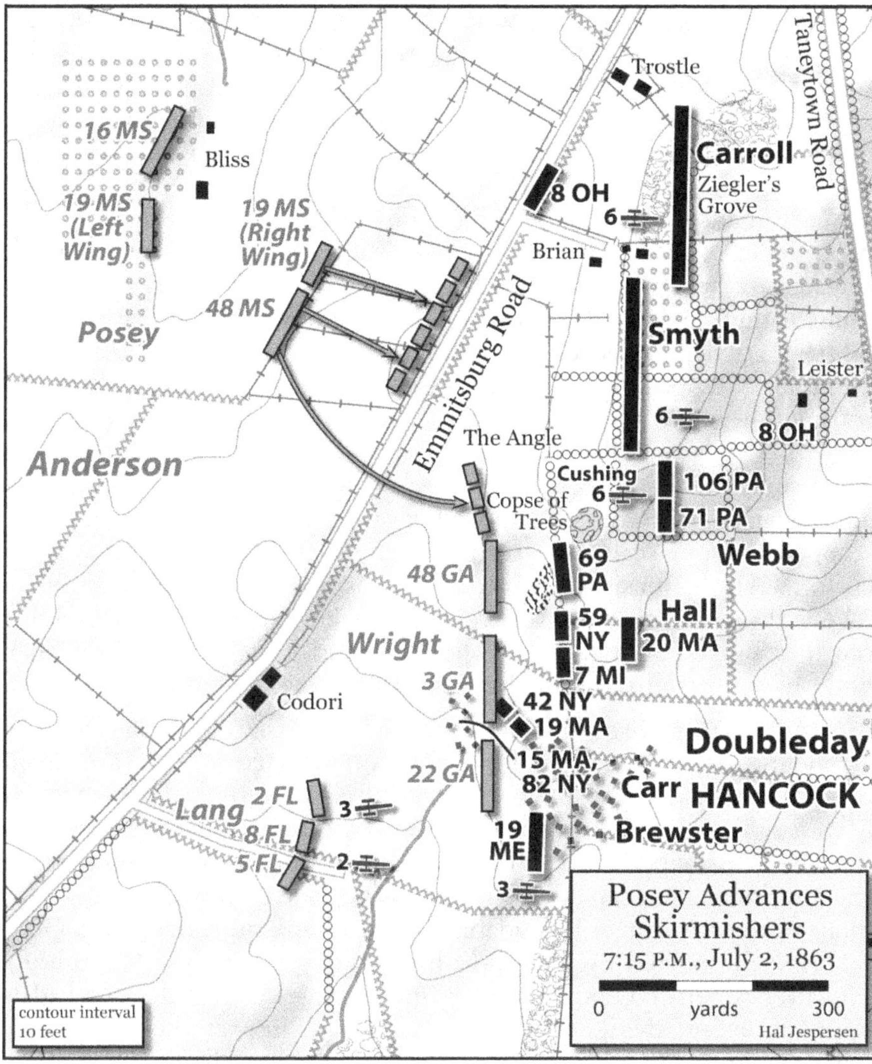

brigade was lying down close behind in the wood, for help to handle the heavy 24-pounder howitzers of Moody's battery. He gave me permission to ask for volunteers, and in a minute I had eight good fellows of whom alas! we buried two that night and sent to the hospital three others mortally or severely wounded.

Losses in Gilbert's Battery, just to the left, were highest on the battlefield with 40 artillerymen killed or wounded of the 75 manning the guns.[4]

4. Ibid., 72; E.P. Alexander, *Fighting for the Confederacy*, Gary W. Gallagher, ed. (Chapel Hill, NC, 1989), 240.

Seething with frustration and anger, Barksdale repeatedly beseeched both McLaws and Longstreet for permission to charge Lt. John K. Bucklyn's Battery E, First Rhode Island Light Artillery— source of his torment— a mere 600 yards away among the Sherfy farm buildings. But Longstreet refused in order to let the en echelon attack to the south fully develop. Finally, sometime after 5 p.m. Cabell's artillery paused and fired three guns in rapid succession — the prearranged signal for Kershaw's brigade, supported by Semmes' brigade, to step off. However, Barksdale and Wofford did not follow immediately on their left being disrupted and delayed, at least in part, by the artillery posted in their front. Barksdale's Mississippians eventually stepped out of the woods and over the stone wall to the rousing cheers of the artillerymen. The four regiments were aligned left to right as follows: 18th Mississippi, 13th Mississippi, 17th Mississippi, 21st Mississippi. Barksdale rode down the line admonishing his colonels to move forward and "swing to the left" and stopped in front of his old regiment, the 13th Mississippi, as was his custom. After a brief address to rouse his troops, Barksdale shouted: "We will proceed to within 75 yards of the entrenchment, withholding our fire. There you will receive the command, Halt! Ready! Fire! Then without command you will charge with the bayonet!'[5]

Only Barksdale's Mississippians charged that day, shoulder to shoulder, in closed order formation on a narrow 350-yard front bounded on the right by the Peach Orchard (targeted by the 21st Mississippi) and on the left by the Sherfy house and barn (targeted by the 18th Mississippi). The 13th Mississippi and 17th Mississippi struck the interior in between. These four Mississippi regiments (totaling 1420 men) were assaulting a formidable line of nine Union regiments (3000 infantry) supported by two-and-a-half artillery batteries (10 cannon). However superior Union numbers did not prevail this day. Barksdale willingly accepted higher initial casualties to impact the Union line with maximum shock value in order to quickly collapse the Federal salient. Although the Union defenders fought valiantly, this Mississippi juggernaut proved irresistible, and the Federals were soon streaming to the rear and northward along the Emmitsburg pike.[6]

Anyone who witnessed "Barksdale's Charge" — whether Confederate or Union— never forgot the spectacle! Typical of these

---

5. J.S. McNeilly, "Barksdale's Mississippi Brigade at Gettysburg," Publications of the Mississippi Historical Society, Vol. XIV, 1914, 235-36; T.M. Scranton, January 24, 1924 to *Memphis Commercial Appeal*, in *Barksdale Family History*, 285-86.
6. Pfanz, *Gettysburg: The Second Day*, 327; Benjamin Grubb Humphreys, Unpublished Autobiography, MDAH, Chapter XIV, 12; Hugh Bicheno, *Gettysburg* (London, UK, 2001), 26, 119.

recollections was a quote by a Union colonel stationed at the Peach Orchard who commented many years later at a Gettysburg Reunion:

> **I am sure it was the grandest charge that was ever seen by mortal man. Nothing we could do seemed to confuse or halt Barksdale's veterans .... You just came on, and on, and on... the carnage you inflicted on us boys in blue would be impossible to detail.... I would like to shake the hand of every member of the Barksdale brigade who is here today, and I now and here pass this encomium upon them, that they are the bravest men I ever met or ever expect to meet.**[7]

On Barksdale's extreme right Colonel Benjamin Humphreys and the 21st Mississippi attacked the 68th Pennsylvania head-on along the east side of the Emmitsburg Road at the Peach Orchard. Simultaneously the 17th Mississippi attacked the right flank of the 68th Pennsylvania and the pressure became overwhelming. Unleashing deadly volleys at close range, and sounding the high-pitched Rebel Yell, the 21st Mississippi surged forward and soon cleared the orchard of Union defenders, capturing Brigadier General Charles Graham who refused to retreat. During this action, Major Isaac Stamps, nephew of Confederate President Jefferson Davis and Colonel Humphreys' son-in-law, was mortally wounded by a bullet to the abdomen. (Stamps died the next day in the field hospital at the John Crawford farm.) Ignoring instructions to "swing to the left," Humphreys and the 21st Mississippi proceeded eastward to clear or capture several Union batteries which could threaten the flank and rear of Barksdale's three remaining regiments turning up the pike. Ultimately the 21st Mississippi crossed over Plum Run but was forced to withdraw thereby abandoning the captured Union cannon.

Simultaneously, when in range, the Mississippians leveled their rifles and delivered a devastating volley into the Union ranks and batteries all along the line. Barksdale's retribution against Bucklyn's Battery was swift and deadly—29 officers and men were killed or wounded along with 40 horses lost (highest within Sickles' Third Corps artillery). Bucklyn himself had two horses shot beneath him and suffered a shrapnel wound to his left shoulder, incapacitating him for six months. He bitterly commented, 'My battery [was] torn and shattered and my brave boys have gone never to return. Curse the Rebs.'[8] Collis' 114th Pennsylvania Zouaves bravely crossed the

7. McNeilly, "Barksdale's Brigade at Gettysburg," 240.
8. Bucklyn to John B. Bachelder, December 31, 1863, *Bachelder Papers: Gettysburg in Their Own Words*, David L. and Audrey J. Ladd, eds. (Dayton, OH, 1994) Vol 1, 72-73.

Emmitsburg pike to confront the 17th Mississippi head-on and enable the battery to escape. This prompted their sister regiments, the 57th Pennsylvania and 105th Pennsylvania, to also cross and take up better firing positions amongst the Sherfy house and farm buildings in order to contest the advance of the 13th Mississippi and 18th Mississippi and protect the right flank of the 114th Pennsylvania.

Barksdale's shrieking Mississippians swept toward the remaining regiments of Graham's brigade which soon "tumbled like dominoes" in the words of historian Stephen Sears. Captain Edward R. Bowen described the collapse of the 114th Pennsylvania: "...as the enemy had already advanced so quickly and in such force as to gain the road and, pouring a murderous fire on our flank, threw the left wing of the regiment on to the right in much confusion ... my men falling in such numbers." The 17th Mississippi closed upon the right of the 13th Mississippi and moved northward along the roadway. Meanwhile the 18th Mississippi cleared the Sherfy barn of Union defenders and overran the Sherfy house so quickly that riflemen from the 57th Pennsylvania who "kept on firing from the windows" after their comrades had withdrawn, were taken prisoner.[9]

The 73rd New York "Fire Zouaves" from Brig. Gen. Andrew A. Humphreys' Second Division had been deployed by Sickles in an open field east of the pike one hundred yards to the rear of the 114th Pennsylvania to support Graham's brigade. Withholding their fire for what must have seemed an eternity until the fleeing Zouaves cleared their front, and facing the onrushing Mississippi brigade, "which came swarming up the slope, yelling like demons and led by Barksdale," Captain Frank E. Moran witnessed these New York firemen unleash "a volley at the [Mississippians] who fell in scores among the dead and wounded Pennsylvanians." Staggered by this "fresh fire," Barksdale's men "waved their flags, cheered, and returned [the] volley." This supporting Union line was soon ordered to fall back, but Captain Moran lay wounded and could not withdraw: "... I waited in a tornado of bullets and shells on my one good leg and my painful eye, until the 13th Mississippi, came over me cheering and firing ... as [they] passed over me a fresh and terrible volley hissed through them from the front ... and scores of [Rebels] went down." One of these casualties was Colonel James C. Carter of the 13th Mississippi who was killed in the roadway— pierced by four rifle balls— while entreating his men to advance into the fields beyond.[10]

9. McNeilly, 238; Stephen W. Sears, *Gettysburg* (Boston, 2003), 299; OR 27/1, 503; Humphreys Autobiography, Chapter XIV, 13; *History of the Fifty-Seventh Pennsylvania Volunteer Infantry* (Kearny, NJ, 1995), 91-92.

10. Captain Frank E. Moran to John B. Bachelder, January 24, 1882, in *Bachelder Papers*, Vol. 2, 771-74; J.C.L. [Pvt. Joseph C. Lloyd] *Daily News*, Meridian, MS, April 13, 1901.

The next Union regiment to the north, the 105th Pennsylvania "Wildcats" (supported by the six Napoleons of Lieutenant Francis Seely's 4th United States Battery positioned at the Trostle Lane) changed front 90 degrees, deployed across the Emmitsburg road, fired, and countercharged, temporarily pushing Barksdale's Mississippians back to the Sherfy House. At this critical moment all the hours of drilling over the past two years under Barksdale's strict discipline now paid huge dividends. Catching their breath, and under the close supervision of their line officers, these Mississippians quickly overlapped both flanks of the 105th Pennsylvania and sent the Union survivors scurrying to the rear. Barksdale's three regiments then advanced northeastward to attack General Humphreys' division, forcing the Union commander to "refuse the flank" to meet them. The 11th New Jersey pivoted ninety degrees to occupy the orchard behind the Klingle farm facing south while Colonel W.R. Brewster's Excelsior Brigade, comprised of the 71st, 72nd, 73rd, 74th and 120th New York regiments, extended this new line southeastward north of the Trostle lane. Simultaneously Brigadier General Cadmus Wilcox's Alabama Brigade and Perry's small Florida Brigade (commanded by Colonel David Lang) of Anderson's Division, Hill's Third Corps, attacked en echelon eastward against Humphreys' main defensive line comprised of Carr's Brigade facing west along the east side of the Emmitsburg Road.

Here occurred some of the most savage fighting of the afternoon attested by the horrendous casualty figures. Humphreys' Division suffered 2092 killed, wounded, or missing (42.5%). It was also likely that Barksdale, conspicuous on horseback, here received two serious wounds to his left leg — fracturing it and resulting in serious blood loss. He reeled in the saddle but doggedly pressed on with his customary exhortation: "Forward, men Forward!" Attacked from front and left flank, with no available reserves, hard-bitten professional soldier Humphreys realized his position was untenable and required a slow deliberate, fighting withdrawal. Humphreys later admitted: "For a moment I thought the day was lost ... so far as I could see, the crest in my rear [Cemetery Ridge] was vacant .... Twenty times did I bring my men to halt and face about." Young staff officer, Jesse Bowman Young, feared the worst: "When our division began to fall back under the awful pressure which was incumbent from three directions it seemed to my distracted boyish vision that the Union was going to pieces, and that Lee was winning a decisive victory on the soil of my native state." It was indeed a very close-run struggle with a razor thin margin of victory in the balance.[11]

11. James A. Hessler & Britt C. Isenberg, *Gettysburg's Peach Orchard—Longstreet, Sickles and the Bloody Fight for the "Commanding Ground"*

Unfortunately for Barksdale and his Mississippians the expected support from Wofford's Georgia Brigade, 1630 strong and directly behind them at the outset, was nowhere to be found. Unbeknownst to Barksdale, when Wofford's brigade had reached the Emmitsburg Road, accompanied by Longstreet, instead of swinging to the left to support Barksdale in accordance with the battleplan, these Georgians advanced directly eastward astride the Millerstown Road (Wheatfield Road) in support of Kershaw's South Carolinians and Semmes' Georgians then attacking the Stony Hill and Wheatfield. Despite Barksdale's repeated requests for reinforcements, with Pickett's Division not yet on the battlefield, there were none available to send. As Humphreys' Union line finally gave way and stampeded to the rear, Colonels Holder and Griffin implored Barksdale to halt and reform his battle line, now ragged after advancing close to a mile under fire. A Mississippian recorded the fiery Mississippian's terse response: "But [Barksdale's] fighting blood was up. 'No' he replied, Crowd them—we have them on the run. Move on your regiments." Another Mississippian later recalled, "these veterans, now covered with dust and blackened with the smoke of battle, with ranks depleted by shot and shell, and faint from exhaustion, responded with cheers to the clarion call of the intrepid Barksdale as he moved bravely on."[12]

"Old Barks" sensed the moment of victory was at hand and shouted, "Brave Mississippians, one more charge and the day is ours!" — the only opposition he could see was a few artillery pieces, completely unsupported by any infantry, some 400 yards to his front. This was the so-called "Plum Run artillery line" frantically cobbled together from battery remnants by Colonel Freeman McGilvery just in the nick of time. Wilcox stated: "From the batteries on the ridge ... grape and canister were poured into our ranks" while a member of the 21st Mississippi recalled: "thinned by the storm which swept down with such terrific fury from the ridge, the advance line staggered and began to waver; the awful crisis of that awful day had come." These guns held the hard charging brigades of Barksdale and Wilcox at bay upwards of thirty minutes — just long enough to permit the arrival of fresh infantry support.[13]

Major General Winfield Scott Hancock, now commanding the entire Union left wing, rushed Colonel George L. Willard's New York

Along the Emmitsburg Road (Eldorado Hills, CA, 2019), 218; Pfanz, *Gettysburg: The Second Day*, 331; Campbell, "The Key to the Entire Situation," 188-89; Jesse Bowman Young, *The Battle of Gettysburg: A Comprehensive Narrative* (New York, 1913), 256-57.

12. McNeilly, 243; Love, "Mississippi At Gettysburg," 32.

13. Campbell, "The Key to the Entire Situation," 196-97; Maj. John J. Hood, "Tribute to Gen. Barksdale," *Confederate Veteran*, November, 1901, 503.

Brigade into the breach, precipitating an encounter for the ages — right out of a Hollywood movie script! Nine months earlier Barksdale's Mississippians had stampeded the untested 126th New York from Maryland Heights earning these Yankees the sobriquet, "Harper's Ferry Cowards." For their part, Barksdale's Mississippi Brigade was desperate to restore their reputation after their thin battle line was pierced by overwhelming Union numbers at Second Fredericksburg (during the Chancellorsville Campaign two months earlier) and was forced to retreat. As Barksdale exhorted his Mississippians onward through the Plum Run Swale to exploit the gap, the 125th and 126th New York regiments rushed forward to meet them head-on. The regimental historian of the 126th New York described the scene:

> **At this critical moment our line wavered, when a voice our boys had heard before was heard cheering on the enemy. It was BARKSDALE! ... A low cry, 'Remember Harper's Ferry!' was heard in our ranks, and swelled into a shout from hundreds of voices .... The venom in that old taunt, 'Harper's Ferry Cowards,' which had so long burned in the veins of that noble regiment, now excited them to fury .... Gen. Barksdale was trying to hold his men, cheering them and swearing, almost frantic with rage.**[14]

Emerging from the bushes these two Union regiments fired a volley and Barksdale fell from his horse with a gaping exit wound in his chest which proved mortal. He died at 2 a.m. next morning at a Union field hospital at the Hummelbaugh house where he was temporarily buried. Willard soon received a horrible shrapnel wound which carried off half of his face. The 'Redemption of the Harper's Ferry Cowards' was complete, and Barksdale's 'most magnificent charge' was finally repulsed. Both reputations had been restored but sadly neither Barksdale nor Willard survived to bask in the glow of their accomplishment.

Barksdale's Mississippi Brigade had paid a horrible price in blood to pierce the Union line and inflict enormous casualties on the enemy. Of the 1420 officers and men who made the charge, 747 were now killed, wounded, or missing. Indeed, Barksdale's brigade suffered the highest casualty rate (49.7%) of any Confederate brigade engaged that afternoon. The three regiments who obeyed Barksdale's command to "swing to the left" each suffered in excess of 50% casualties with the

14. A.M. Willson, *Disaster, Struggle, Triumph: The Adventures of 1000 Boys In Blue* (Albany, NY, 1870), 168-69; Maj. Charles A. Richardson to John B. Bacheldelder, May 8, 1868 in *Bachelder Papers*, Vol. 1, 338-40; Ibid., August 18. 1867, 315-17.

17th Mississippi topping out at 57.7%. The 21st Mississippi who attacked eastward through the Peach Orchard suffered comparatively less at a 33.8% loss. The regimental command structure within Barksdale's Mississippi Brigade had been shattered — Colonel Carter (13th Mississippi) was killed and Colonels Holder (17th Mississippi) and Griffin (18th Mississippi) were both wounded, while only Colonel Humphreys of the 21st Mississippi survived unscathed. Private Moore of the 17th Mississippi recorded his assessment of the day's action in his diary, and spoke for the entire brigade:

> **Drove them before us for 1-1/4 miles but were forced to fall back for** *lack of support* **(emphasis added). Every man acted the hero. Miss has lost many of her best and bravest sons. How thankful should all be to God who have escaped. OH! the horrors of war.**[15]

As the Confederate en echelon attack progressed northward the Georgia Brigade of Brigadier General Ambrose 'Rans' Wright attacked eastward across an open field on the left flank of Colonel Lang's Floridians and succeeded in clearing the Emmitsburg Road of Union infantry and captured two Union batteries nearby. Noticing no support from Posey's Mississippi Brigade on his left, Wright sent an immediate request to Anderson for Posey and Mahone to advance their brigades, but still pushed on through the Cordori farm to the main Union line on Cemetery Ridge. Here these brave Georgians captured several additional artillery pieces but, with Lang's Floridians streaming to the rear on their right, and Posey and Mahone remaining absent on their left, Wright was forced to retire, abandoning the captured cannons. The furthest location that Wright's Georgians actually reached remains in doubt.

Anderson's single division attacked along a front of equal length as the one Longstreet advanced against with his two divisions (Hood and McLaws). With half as many men Anderson's assault was accordingly only one brigade deep (same as Barksdale). Lacking depth and support, the gains could not be held by Wright. Also, unlike Longstreet who commanded from the front lines and could readily witness the action, Anderson maintained his headquarters in rear of Seminary Ridge out of view. Sickly A.P. Hill remained behind Seminary Ridge in the vicinity of Lee's headquarters throughout. Accordingly, command and control were lacking, and communication was significantly hampered. The end result was that Posey advanced only two of his four regiments

15. Hessler & Isenberg, *Gettysburg's Peach Orchard*, 240-41; *Robert A. Moore War Diary, A Life for the Confederacy*, James W. Silver, ed., (Jackson, TN, 1959), 153.

forward as skirmishers, and Mahone refused to advance his 1,542-man brigade (second largest in the division) at all. In fact, Gettysburg authority Edwin B. Coddington estimates that Anderson deployed only 4,300 officers and men of the 8,300 available![16]

Posey's Brigade was by far the least participatory of the three Mississippi brigades during the battle of Gettysburg, as borne out by their extremely low casualty figures — 12 killed, 71 wounded, totaling 83 out of 1,150 present (7.2%). They arrived on the battlefield late on the afternoon of July 1st eager to engage the enemy, but were ordered to halt and stack arms in the vicinity of Herr Ridge. Lee later explained that, with the scope and final outcome of the battle still in doubt, he held back Anderson's sole remaining unengaged division since "a reserve in case of disaster was necessary." During the morning of July 2nd Posey's brigade assumed a position in the open fields between McMillan and Spangler's woods in support of Major William Pegram's Artillery battalion on Seminary Ridge. Around 10 a.m. members of the 16th Mississippi and 19th Mississippi became embroiled in the bitter and prolonged skirmishing at the Bliss farm buildings midway between the main Union and Confederate lines. The Bliss barn was an ideal sharpshooter position and changed hands several times that day as both sides stubbornly fought to occupy it. Late that afternoon, while the 16th Mississippi (now reinforced by the 12th Mississippi) still stubbornly skirmished at the Bliss farm, General Posey received an order to support General Wright's attack on his right. Amidst the chaotic situation there, and without support on his right, Posey advanced portions of the 48th Mississippi and right wing of the 19th Mississippi beyond the Bliss farmstead and deployed them as skirmishers ? some of whom mingled with Wright's Georgians and advanced in a disorganized manner. These two regiments accounted for the majority of casualties within Posey's brigade—10 killed and 47 wounded. As darkness fell and the Confederate assault sputtered out, Posey's Mississippi Brigade was ordered to return to their original position supporting Pegram's batteries on Seminary Ridge. These Mississippians remained there throughout the following day without engaging the enemy except for additional skirmishing by the 12th Mississippi at the Bliss farm in the morning and early afternoon when Union troops, under direct orders of Brigadier General Alexander Hays, finally torched the Bliss barn and house.[17]

16. *OR* 27/2, 613-14, 621-24, 631-34; Edwin B. Coddington, *The Gettysburg Campaign: A Study In Command* (New York, 1968), 420-26; n84, 760-61.

17. Posey Mississippi Brigade monument at GNMP; Eric A. Campbell, "Sacrificed to the bad management ... of others": Richard H. Anderson's Division at the Battle of Gettysburg, The Army of Northern Virginia in the

As for Davis' Mississippi Brigade, bloodied on the first day, these Mississippians were held in reserve on July 2nd licking their wounds and gathering arms and accoutrements behind Confederate lines. In the evening they were joined by the 11th Mississippi from Cashtown and moved up to Seminary Ridge, halting in McMillan's Woods.[18]

* * *

PUBLICATIONS OF THE MISSISSIPPI HISTORICAL SOCIETY
VOL. XIV, 1914, P.238
"BARKSDALE'S MISSISSIPPI BRIGADE AT GETTYSBURG,"
J.S. MCNEILLY

CLARION LEDGER
JUNE 11, 1891

He Wanted a Wagon.

Commercial Herald: The reunion at Jackson has revived many old Confederate stories which have been rehearsed by those who witnessed them when they occurred. One of these is told of Gen. Wm. Barksdale and happened only the day before he was killed, the second day's fight at Gettysburg. Gen. Barksdale had made his brigade an eloquent speech that afternoon exhorting them to display their accustomed gallantry in the coming fight, and concluding with the words "if any one of you is afraid to let him come to me and tell me so, and I will give him a wagon to take care of." A brigade that had already been in so many engagements was not likely to have been in it who wanted to be teamsters when fighting was to be done, but that night, much to the commander's rage, a private, prompted by the spirit of deviltry that frequently accompanies daring, called at General's quarters, sought an interview with him and asked for the proffered wagon, but got instead a raking down, fore and aft, that doubtless cured him of all his mischievous propensities forever.

JUDGE J.B. BOOTH
21ST MISSISSIPPI, COMPANY F
POST-WAR LETTER

I remember vividly the effects of the first shot that came from the battery in our front, that called forth General Barksdale's request to General Longstreet to allow him to order in his brigade and take the battery. The shell exploded in the ranks of my company, near me. J.T. Worley was killed and Capt. H.H. Simmons, John H. Thompson and

Gettysburg Campaign 1998, Gettysburg National Military Park Seminar 1998 (Gettysburg, GNMP, 1999); Elwood W. Christ, *"Over a Wide, Hot ... Crimson Plain": The Struggle for the Bliss Farm at Gettysburg, July 2 and 3rd 1863* (Baltimore, MD, 1992), 20-21; 27-31; 38-40; 47; 74-76.

18. Sgt. Augustus Vairin Diary, July 2, 1863; Stubbs, *Duty*Honor*Valor*, 414.

John T. Neely each lost a leg but survived the war. John T. Neely died a few days ago (July 5th) at his home in Charleston, Miss. By the same shot, there were other casualties but I did not recall the names and companies of those killed and wounded. Captain Stamps was killed in three feet of me soon after the brigade went forward. We were just entering the peach orchard when he was stricken down. [Ed. Note: Stamps was mortally wounded in the bowels and died the next day surrounded by his men at the brigade hospital at the John Crawford farm]

* * *

CONFEDERATE VETERAN,
JULY 1893

CAPT. W. GART JOHNSON
18TH MISSISSIPPI, COMPANY C
ORLANDO, FLA., JUNE 26, 1893

BARKSDALE HUMPHREY'S MISSISSIPPI BRIGADE

[Ed. Note: The following excerpts describe the character of Brig. Gen. William Barksdale and Colonel Benjamin Grubb Humphreys, 21st Mississippi as observed by an officer who served with them]

Barksdale the ranking colonel ... took immediate command, was promoted to the rank of general in a few days, and continued in command till he fell, leading his men at Gettysburg.... He was a man of whom it could be truthfully said, "Bold as a lion, yet gentle as a lamb." He was not a military man, but was a pure type of genuine southern chivalry, a southern gentleman of the old school. Quick to resent and as quick to forgive, quick to punish disobedience in a subordinate, and as quick to ask forgiveness. Just as far removed from military hauteur as one could imagine. I cite one instance: We were in camp, it was one summer evening. Gen. Cobb, of Georgia, his old friend and former fellow Congressman, had dined with him. We were drilling, when the two Generals, arm in arm, coats off, came walking out to look on. It reminded me very forcibly of two farmers in antebellum days taking an afternoon stroll through the farm to look at the crop. We loved Gen. Barksdale, because we knew he was proud of us, and would do any thing in his power for our welfare. No truer patriot ever fell on the field of battle.

Gen. Humphreys was a West Pointer, was there at the same time with Mr. Davis, but unlike him, he chose a more peaceful calling, that of a planter in the rich bottom lands of the Mississippi, where he could enjoy the peace and quiet of home life, and indulge in his favorite sport

Captain W. Gart Johnson of the 18th Mississippi Infantry

(Confederate Veteran Magazine)

of hunting deer and bear. For the rank he held, and as a commander of infantry, I do not think he had a superior in either army. He possessed all the qualities, both natural and acquired. He won the love of both officers and men by his great kindness. He won their unbounded confidence by his coolness and ability under the most trying circumstances. He was approachable on all occasions. His officers and men obeyed him implicitly, not because they recognized his right to command them, but because to comply with an order from him was the right thing to do.

[Ed. Note: Humphreys was expelled from West Point during the Christmas riot in 1827 and did not graduate. He later lamented: "I have ever regretted my error and regard it as one of the greatest misfortunes of my life.]

* * *

CONFEDERATE VETERAN,
MAY 1906, P.217

J.W. DUKE
CO. C, 17TH MISSISSIPPI REGIMENT, TROY TENN.

MISSISSIPPIANS AT GETTYSBURG

I was a member of Company C, 17th Mississippi Regiment, Barksdale's brigade, and will give you my experience in the battle of Gettysburg, Pa., July 2, 1863. On the 1st of July we marched all day until late in the night, then halted for a few hours' rest. At four o'clock on July 2 we were called to arms, and a detail of ten men from each company drew twenty rounds of extra cartridges for the bloody fray. My brother and I were on this detail. As we were going to the ordnance wagon my brother said he wanted me to write home after the fight. I replied that it was his time to write, as I wrote last. It was yet dark, but I saw tears in his eyes. He said to me: "Something is going to happen today." I think my brother had a presentiment that he would be shot. In the first charge in the famous old peach orchard he was wounded, but I did not know of it until night. Our adjutant, Lieut. Ashley, told me where he fell. I hunted until I found him. His first words to me were: "Thank God! My prayers are answered. I have asked him to take me in place of you, as I am prepared and you are not."

Comrades, this is the first time that I ever weakened. I promised him there that I would live a better life in the future, but it was many years before I did. His leg was amputated, but he lived several weeks. Gangrene in the wound caused his death. He died happy, so our chaplain wrote me. I am trying to meet him over the river.

As a prisoner, I was sent to Rock Island. I enjoyed Miss Kate E. Perry's article in the February VETERAN, as I had many favors from her while there. I knew all of the prisoners who made their escape by her aid.

Recurring to the battle of Gettysburg, I will further say that on the morning of the 2nd, after we were marched in the enemy's front, Capt. Cherry, of Company C, was ordered to the skirmish line and to lie down. While lying there Gen. Longstreet came up and told the Capitan to send two men without guns or cartridge boxes to a house with paled fences about the yard and garden and knock off the palings, so that he could make the attack. Capt. Cherry told the orderly to call the first two men on detail. He called, and they failed to get up, so he called the next two, and they also failed. The captain became enraged and said: "I will make the detail." He called out: "Jim Duke and Woods Mears, they will go." We got up and went. I said to Woods as we went: "We will be killed." We knocked off the palings, and they never bothered us. There were three lines of infantry in about fifty steps, and two hundred yards farther there were six pieces of artillery. When we returned, Gen. Longstreet was still standing there. He said to me: "Buddie, what did you see there?" I told in a few words. I was dreading Gettysburg heights and I said: "General, do you think we can take those heights?" He replied, "I don't know, do you?" and then he said, "This is not my fight." That excited me, for I had great confidence in Longstreet.

In a few moments we were ordered forward and to hold our fire. We killed or captured everything in front of us until we were near their works. There Gen. Barksdale was shot. As he fell from his horse he told Col. Humphreys to take charge of the brigade and move it to the rear. So ended the second day's fight at Gettysburg.

* * *

PUBLICATIONS OF THE MISSISSIPPI HISTORICAL SOCIETY
VOL. XIV, 1914, P.240
"BARKSDALE'S MISSISSIPPI BRIGADE AT GETTYSBURG,"
J.S. MCNEILLY

JNO. S. HENLEY
17TH MISSISSIPPI REGIMENT
ABERDEEN EXAMINER

GETTYSBURG

Colonel Worthen commanded a regiment stationed along the old Emmetsburg road in front of the peach orchard, supporting ten guns parked on an elevation near the orchard; this witness told Mr. Henley the following brief yet stirring story of what he saw Barksdale and his men do.

"Yonder," said he, "Barksdale formed his line in that clump of woods, between us and beneath us under heavy fire of the ten guns that we were supporting; between that clump of woods, Barksdale and his brigade had to pass over open fields to get to us; while he was forming his men in line of battle, the missiles from the ten guns tore the limbs off of the trees and plowed gaps through his men, yet in battle line they stood and as soon formed, they left the woods to traverse the open field intervening. When our sharpshooters and pickets opened fire on you, as you left the woods, besides the mischief done by the sharpshooters and the pickets, the ten guns were hurling missiles of death into your ranks and swept gaps through them all the way across the field, and when a solid shot tore a gap in your ranks, it was instantly closed up, and the brigade came on in almost perfect line."

And said the brave Colonel," I am sure it was the grandest charge that was ever seen by mortal man. Nothing we could do seemed to confuse or halt Barksdale's veterans. We federals had five regiments fronting Barksdale's small brigade, and these were supported by two additional regiments stationed just behind the five on the embankment, east of the peach orchard, but nothing daunted Barksdale and his men. You just came on, and on, and on, and when you came in gun shot of us, the carnage you inflicted on us boys in blue would be impossible to detail. Before we could get them to the rear the ten guns were captured, the eighty horses killed and you then drove our five regiments from the field, ran over two lines of reserves as you did the five regiments in front, and carried everything before you with cyclonic force. I would like to shake the hand of every member of the Barksdale brigade who is here today, and I now and here pass this encomium upon them, that they are the bravest men I ever met or ever expect to meet."

* * *

Account Found in 17th Mississippi Folder,
Gettysburg National Military Park

Private John S. Henley, 17th MS, Company A

On The Way To Gettysburg [July 2nd]

We were ordered to attack at once. Hood's Division on our right and Wilcox's Brigade on or left. The Federals were strongly posted on the high ground at the famous Peach Orchard. They had a line of battle on the Hagerstown [Emmitsburg] Road, which had been dug out giving them some protection. They had another line across the Peach orchard, and still another line back of the orchard. We got our line formed, and opened on them about one o'clock. The fire was heavy, and all the high ground about the peach orchard was covered with artillery. We shelled them about an hour. They killed so many of our artillerymen that some of the infantrymen had to go and help them handle the guns. At about 2 o'clock we moved forward. We had to go one mile through the field, with all of their artillery firing on us. We went in perfect line. They would knock large gaps in our line. Then we would fill up the gaps and move on. When we did get to the Federals, I quote the words of Col. Worthen, who commanded a regiment of federals in the orchard, "We liked to have killed them all, before they could get away." He told me years afterward that he only had six men left, who were able to perform duty in the regiment. That regiment began to run, and we commenced firing on them, and they ran in crowds. You could not shoot without hitting two or three of them. We went through the Peach orchard, and lost many of our men. A Pennsylvania regiment was posted behind an embankment, and they killed lots of our boys....General Barksdale was killed soon after we passed out of the Peach Orchard. It was in a field, no trees about. Col. Holder, commander of the 17th Mississippi, was wounded in the stomach. His entrails were shot out, and he held them in until he got to a hospital.... I will say that Col. Holder was the best man who ever commanded the 17th Mississippi Regiment. He had the respect of the entire regiment. The wheat field that we had fought through was a beautiful sight before the battle, the heads were full and just turning yellow.... My company had 42 men when we went into the battle, and we had 21 killed and wounded.

* * *

CONFEDERATE VETERAN
MAY 1909, P. 229

JUDGE GEORGE CLARK, WACO TEXAS

WILCOX'S ALABAMA BRIGADE AT GETTYSBURG

About sunrise the next morning (July 2nd) the brigade formed line and moved to the left and east, and upon reaching the proper point fronted and began to move forward in line to the position assigned us for the battle. The 11th Alabama occupied the left of the line, and after moving forward a short distance entered a valley and an open wheat field, and when about halfway across the field were fired into by a brigade of federal sharpshooters in the woods on our right and rear which produced some confusion and a retreat back to the fence, so as to escape the fire from the rear. But just at this time the 10th Alabama came up on our right and immediately opposite the federals in the woods, and after a brief musketry drove the enemy back and uncovered the right flank of the 11th Alabama, thus enabling the brigade to move forward in line and take position, which was done at another fence across the field.

Here we remained almost the entire day and until 4 P.M. The sun was fiercely hot, and there was no shade or other protection for the men. Here they sweltered and sweated and swore until about four in the afternoon, when the engagement began on the right.

Captain George W. Clark of the 11th Alabama Infantry

(Texas State Preservation Board)

Our brigade commander during the morning took occasion to explain to the officers and men the general plan of the battle, in so far as our immediate front was concerned, stating that the movement forward would be by echelon, beginning with the right of Longstreet's Corps and extending to the left as each brigade came into action, and that, owing to our situation, the Alabama Brigade at the proper time would move by the left flank rapidly, so as to give Barksdale's Mississippi Brigade, which would be on our immediate right, room to move forward in proper line.

Thus matters stood until about 4 P.M., when the thunder of cannon up on the right announced the beginning of the action. As Longstreet's brigades came into action the roar of the cannon was accompanied by the rattle of musketry, mingled with the yells of our boys as they moved forward on the run, and the scene was grand and terrific. As the fire and clamor approached the Alabama Brigade Barksdale threw forward his Mississippians in an unbroken line in the most magnificent charge I witnessed during the war, and led by the gallant Barksdale, who seemed to be fifty yards in front of his brave boys. The scene was grand beyond description.

* * *

Times Dispatch
Richmond, VA
July 13, 1913

Private James O. Matthews
21st Mississippi Infantry

Barksdale's Mississippi Brigade At Gettysburg

The fiftieth anniversary of the battle of Gettysburg, and the interest manifested in the celebration, appears to the writer of this sketch to be a fitting time for consummation of a purpose of many years, of writing the story of Barksdale's Brigade at Gettysburg, as he witnessed and remembers it; and under the enlightenment of record reports and authoritative descriptions. It is only regretted that one more gifted, has not told the tale of deeds worthy of the pen of Walter Scott or Hugo. In default of a better annalist the self-imposed task is taken up under the inspiration of duty to the brigade organization and numbers, with a tincture of natural pride in having been a sharer, as a member of the Twenty-first Mississippi, to what was far and away the most thrilling, successful and altogether matchless episode of that greatest and perhaps most momentous of the battles of the war, largely eclipsed though it has been in historic narratives by the dire consequences of that terrible tragedy, the immortal and fatal forlorn hope Confederate march on the next day, into the yawning jaws of slaughter and defeat. Nevertheless, the superlatives used upon Barksdale's charge will be justified by quoted testimony competent witnesses of other commands. Northern and Southern, descriptive of that grand and resistless rush of 1,500 Mississippians upon and over Peach Orchard Hill.

After a wearing march from near Chambersburg, the brigade, a part of General Longstreet's Corps went into camp about midnight six miles from Gettysburg, after the first day's battle. They reached the front the next morning about 9 or 10 o'clock, on the 2d of July, and moved to the right of the line. In order of battle as made by General Lee and entrusted to General Longstreet, the assault was to be opened by the extreme right brigade, the others of the two divisions of Hood and McLaws to join in successively. To Barksdale, on the extreme left, was allotted the taking of Peach Orchard Hill., "a piece of elevated ground," wrote General Longstreet, 'that General Lee desired me to take and hold for artillery."

The brigade was placed in battle line - formed with the Twenty-first, Colonel B. G. Humphreys, on the right, then the Seventeenth, W. D. Holder, next the Thirteenth, Colonel J. W. Carter, with the Eighteenth Colonel Thomas M. Griffin, on the left - under the crown of low ridge, five or six hundred yards distant from the position of assault. Open fields, fences and scattered from houses lay between. Parallel with our line, and at the base of the Peach Orchard Hill, ran the Emmittsburg Road, between two high rail fences. Farther to the left a picket fence lay beyond the road. The noise of hard battle, where Hood's Division opened the fight, was heard far to our right soon after 4 o'clock. The resistance was more obstinate and the wave of attack was longer in reaching us than calculated. The brigades of Kershaw and Semmes, South Carolina and Georgia, of our division, McLaws, went in soon after 5 o'clock on our immediate right, but hidden from view by rolling ground.

While waiting their turn, Barksdale's men lay under fire of artillery and infantry in their front, which they were not allowed to return for an hour or more. Being pretty well covered there were only a few casualties. Never was a body of soldiers fuller of the spirit of fight and the confidence of victory. This was betokened by their conversation, in groups, under such shade as they could find from a hot sun. Some strolled down to the little stream in their rear, where canteens were filled. Others crossed over and broke off great branches from the numerous cheery trees, which were in full bearing. Relief and diversion came when a dozen or more guns were unlimbered in the depression behind us and quietly rolled by hand to our infantry line. As soon as they were placed they opened up in concert and with a din that was deafening. As fast as the gunner could load they concentrated a fire on the Peach Orchard, which was both terribly destructive and demoralizing. This was kept up for half an hour, though the Union guns were silenced after the first discharges. When the order was given by the battery commanders to cease firing, every man in the brigade knew instantly that "our turn" had come. The scenes and events as they moved thereafter were charged with intensely dramatic and never to be forgotten incidence. General Barksdale's appearance, riding rapidly along the rear of the line, was the signal to the respective regimental commanders to get alert.

Noted for excelling in the sound fullness of the word of command, never before did Colonel Humphreys shout "attention" with such imperative insistence. Like an electric shock it brought every man of his regiment up standing. At the same time General Barksdale turned the right of the line and rode as rapidly down the front, to take his position at the head of his old regiment, the Thirteenth. As he turned he called out to Colonel Humphreys the words which dictated the direction of the attack – "swing to the left."

Next came the ringing command – "Double quick, charge," and at top speed, yelling at the top voice, without firing a shot, the brigade sped swiftly across the field and literally swift across the field and literally rushed the goal. Our men began to drop as soon as they came to attention, and were well peppered in crossing the distance to the enemy. The Twenty-first struck and flanked the Peach Orchard angle. Our left regiment, the Eighteenth, breasted a hot fire from a large brick barn – converted into a fortress by a Zouave regiment of Graham's Brigade – which they captured and burned. The Thirteenth and Seventeenth swept the line between these two salients. All met with stiff resistance. But when the blue coats saw us swarming over the fence and across the Emmitsburg Road, without pausing, they began to "back out," But they fought back bravely, retiring slowly until the firing was at close quarters, when the retreat became a rout in which our men took heavy toll for the losses inflicted on them. So hot was our fire that many of the enemy hid behind great boulders, which were numerous and as our charging line passed over them surrendered and were sent to the rear. General Graham, with becoming courage, rode out of the orchard behind his men – his horse was shot down and he was captured. Several guns and caissons were abandoned in the orchard, others were captured later.

The Peach Orchard line the brigade carried was held by Graham's Brigade of six Pennsylvania regiments of Birney's Division. Measuring the importance attached by the enemy to the position, other regiments were crowded into position; quoting from Colonel Livermore and the record, one regiment each of Ward's and De Trobriand's Brigades of Birney's Division, one of Burling's Brigade, and "when Barksdale appeared two regiments of Brewster's Brigade" of Humphrey's Division of the Third Corps. That is, the position which Barksdale assailed with four regiments, was held by eleven regiments, all veterans, all under General Graham's command, and two batteries of artillery. The encounter resulted as stated.

In a Gettysburg article published some years ago in the Confederate Veteran by Judge George Clark, of Waco, and a veteran of Wilcox's Brigade was the following reference to Barksdale's Brigade:

"Our brigade commander during the morning took occasion to explain to the officers the general plan of the battle, in so far as our immediate front was concerned, stating that the movement forward would be by echelon, beginning with the right of Longstreet's Corps and extending to the left as each brigade came into action; and that, owing to our situation, the Alabama Brigade at the proper time would move by the left flank rapidly, so as to giver Barksdale's Mississippi Brigade, which would be on our immediate right, room to move forward in proper time.

"Thus matters stood until about 4 P. M., when the thunder of cannon up on the right announced the beginning of the action. As Longstreet's brigades came into action the roar of cannon was accompanied by the rattle of musketry, mingled with the yells of our boys as they moved forward on the run, and the scene was grand and terrific. As the fire and the clamor approached the Alabama Brigade, Barksdale threw forward his Mississippians in an unbroken line in the most magnificent charge I witnessed during the war, and led by the gallant Barksdale, who seemed to be fifty yards in front of his brave boys. The scene was grand beyond description.

The order was then given our brigade to move rapidly by the left flank, and the movement was made a full speed until space was cleared sufficient for the Mississippians, and then with right face Wilcox's Brigade moved forward to the assault."

In a paper on Gettysburg, read by General McLaws - who commanded the division of which Barksdale's Brigade was a part - in 1878, before the Historical Society of Georgia, he made special mention of his Mississippians. He said, "Barksdale had been exceedingly impatient for the order to advance, and his enthusiasm was shared in by his command. Barksdale was standing in front ready to give the word and to lead. He was not far from me; and so soon as it was signified to me I sent my aid-de-camp, Captain G. B. Lamar, Jr., to carry the order to General Barksdale, and the result I express in Captain Lamar's words:

"I had witnessed many charges marked in every way by unflinching gallantry, indeed, I had the honor of participating when in the line with the First Georgia Regulars, but I never saw anything to equal the dash and heroism of the Mississippians. You remember how anxious General Barksdale was to attack the enemy, and his eagerness was participated in by all of his officers and men, and when I carried him the order to advance his face was radiant with joy. He was in front of his brigade, hat off, and his long, white hair reminded me of the white plume of Navarre.

"I saw him as far as the eye cold follow, still ahead of his men, leading them on. The result you know. You remember the picket fence in front of the brigade? I was anxious to see how they would get over and around it. When they reached it, the fence disappeared as if by magic, and the slaughter on the other side was terrible. Barksdale, gallantly leading his men in the terrific fight, fell, mortally wounded. The last words of that ardent patriot to fall on the ears of one of his countrymen were "I am killed. Tell my wife and children I died fighting at my post."

The "dash and heroism of the Mississippians" won praise quite as high from the brave but beaten adversaries. As our onrushing line sped down the slope of Peach Orchard Hill many of the men were outstripped and left behind as prisoners. The following incident is told from memory: A tall, fine looking Pennsylvanian insisted on shaking hands as he said "with one of the men who had made the most splendid charge of the war."

Barksdale's success in carrying so strong a position, held by more than double his force, was promoted by a splendid artillery support. The report of General E. P. Alexander, who commanded the artillery engaged against the Peach Orchard, is here quoted:

"About 4 P. M., I placed five batteries in position in action against a heavy artillery and infantry fire of the enemy, about 500 yards distant in a peach orchard on the Emmittsburg Pike. After a spirited engagement of a half hour before the enemy's guns were silenced and the position was silenced and the position was immediately carried by the infantry and the enemy fell back to his position on the mountain where our infantry gallantly pursued him. . . . The sum total of my losses were killed, nineteen; wounded, 114. There were also two killed and three wounded of a detachment of eight gallant Mississippians at Captain Moody's guns, who volunteered to help maneuver them on very difficult ground."

Alexander's guns followed closely upon Barksdale's charge, occupied the Peach Orchard Hill, which we had carried as ordered by General Lee. From this commanding position they continued to render effective support to the onward movement. Barksdale changing direction, moved directly on the flank of the enemy's line beyond where he had smashed it. His Twenty-first was the brigades of Kershaw and Semmes, who were to fill in between Hood's left and Barksdale's right. Thus instead of having a charging column to connect with on his right. Barksdale was necessitated to go up against the Peach Orchard with his right "in the air." Not until night had closed the battle did we connect with the other brigades of the division, which had been held at bay until Barksdale "broke in the Peach Orchard angle," and eased the way for the troops on the right to move forward. Of the relief afforded by Barksdale's success. Longstreet wrote: "Kershaw, joined by Semmes, responded, and Hood's men feeling the impulsion of relief, resumed their bold fight, and finally the enemy's whole line was broken throughout its length."

After clearing up the Peach Orchard Hill, for the artillery to occupy, Barksdale changed front to assail the enemy's line to our left. In this movement his left was joined on by Wilcox's Brigade of Anderson's Division. But after together breaking up and dislodging Humphrey's Division, connection was lost, too. In his Gettysburg writings General Longstreet thus refers to the charge of Barksdale and his brigade on the Peach Orchard position, and the unfortunate separation from Wilcox:

"Barksdale's Brigade was in front of a battery about 600 yards off. He appealed for permission to charge and capture it, but was told to wait. . . . I rode to McLaws found him ready and Barksdale chafing in his wait for the order to secure the battery in his front... After additional caution McLaws ordered Barksdale in.

With glorious bearing he sprang to his work, overriding obstacles and danger. Without a pause to deliver a shot, he had the battery. . . . My men charged with great spirit and dislodged with great spirit and dislodged the enemy from the Peach Orchard wit but little delay, though they fought stubbornly. We were then on the crest of Seminary Ridge. The artillery was brought forward and put its position at the Peach Orchard. The infantry swept down the slope and soon reached the nearby ground that lay between Seminary and Cemetery Ridges, fighting the enemy for every foot of ground and against overwhelming odds. Nothing could stop my men, and they carried their heroic charge up Cemetery Ridge... Touching the failure of supporting brigades of Anderson's Division to cover McLaws (Barksdale's) left flank as directed, there is little to be said. Those brigades acted gallantly, but went astray early in the action. General Anderson in his report, says: "A strong fire was poured on our right flank, which had been detached from McLaws's, they were exposed to attack in front and on both flanks, and compelled to retire." Longstreet is again quoted: "This drew fire Anderson's Brigade of direction. Wilcox from support of Barksdale's left General Humphreys seeing the opportunity, rallied such of his troops as he could, and reinforced by Willard's Brigade of Hancock's Corps, came against Barksdale's flank. But the latter moved bravely on, as guiding spirit of the battle."

Before Barksdale "moved on" to grapple with Willard's fresh brigade, led by General Hancock, he was urged by Colonels Holder and Griffin to halt and reform his men, already greatly reduced by two successful but severe and bloody struggles. But his fighting blood was up. "No" he replied. "Crowd them – we have them on the run. Move on your regiments." In the ensuing conflict Willard was killed, and Barksdale mortally wounded. In retiring, which they did fighting and in order, both Holder and Griffin were disabled by severe wounds. Lieutenant-Colonel Fiser, of the seventeenth, was wounded. Lieutenant-Colonel Luse, of the Eighteenth, captured. Colonel Carter and Lieutenant-Colonel Bradley of the Thirteenth and Captain Stamps, acting major of the Twenty-first, had already been killed. To more fully and clearly explain the brilliant isolated success of Barksdale's Brigade at Gettysburg, rendered fruitless, alas! by its isolation, the Century contribution of General H.J. Hunt, commander-in-chief of Meade's artillery, is quoted:

"The breaking in of the Peach Orchard angle exposed the flanks of batteries on its crest, which retired firing. Many guns of different batteries had to be abandoned; all lost heavily. Bigelow's Ninth Massachusetts Battery made a stand close to the Trostle house in the corner of the field. Although much cut up, he was directed to hold that point at all hazards until a line of artillery cold be formed beyond Plum Run. Finally some twenty-five guns formed a solid mass, which, unsupported by infantry, held this part of the line until the abandoned guns could be brought off. When, after accomplishing its purpose, all that was left of Bigelow's Battery was withdrawn, it was closely pressed by Colonel Humphrey's Twenty-first Mississippi Regiment, the only regiment which succeeded in crossing Plum Run. His men had entered the battery and fought hand to hand with the cannoneers; one was killed while trying to spike a gun, and another knocked down with a handspike while trying to drag off a prisoner. Of 104 men the battery took into action, twenty-eight were killed and wounded, two missing. Of four battery officers, one was killed, another mortally wounded and Captain Bigelow was severely wounded. As the battery had sacrificed itself for the safety of the line, its work is specially mentioned. When Sickles was wounded General Meade directed General Hancock to take command of the Third Corps as well as his own corps. About 7:15 the field was in a critical condition. Birney's Division was now broken up, Humphrey's Division was slowly falling back. Anderson's (Confederate) Division was advancing. On its right Barksdale's Brigade, was except the Twenty-first Mississippi, was held in check only by McGilvery's Artillery, to whose support Hancock now brought up Willard's Brigade of the Second Corps. Willard charged Barksdale's Brigade and drove it back nearly to the Emmitsburg Road, where he himself was repulsed by a heavy artillery and infantry fire, and fell back to his former position. In this affair Willard was killed and Barksdale mortally wounded. Meanwhile the Twenty-first Mississippi crossed Plum Run from the neighborhood of the Trostle house and drove out the men of Watson's Battery, Watson being severely wounded.

The Gettysburg chapters of the "Story of the Civil War," by Colonel W. R. Livermore, U. S. A., a full and detailed history of the great battle, and on the whole an accurate and fair statement of the actions of the different commands, compiled from the official records of both armies, is quoted:

"At 6:20 Barksdale at last advanced on the Peach Orchard.... All Federal batteries except Bigelow's then withdrew, Graham's men rallied at the edge of the orchard, but were obliged to yield. As soon as the apex of Sickle's line was broken, both wings collapsed.

(Author's Note: One of Ward's regiments and one of DeTrobrand's had been detached from their brigades and formed at the southern edge of the orchard. One of Burling's regiments formed on the cross roads near Want's house; the others remained in reserve. All were under Graham's command. When Barksdale appeared, two more regiments from Brewster's Brigade went successively to him.)

"Barksdale followed Graham's men, who fell back towards Humphries.... The retreat of Graham's men exposed the left of Humphrey's Division; and Barksdale veered off to the left to take it in flank. Brewster, Humphrey's left brigade, exposed to an enfilading fire, fell slowly back to the right and rear with great loss. At about 6: 30 P. M., Wilcox, Perry and Wright (of Anderson's Division, Hill's corps) on Barksdale's left, began to advance in succession. (Author's Note: Seeley's and Turnbull's batteries fired on Barksdale and Wilcox as they advanced.) Birney ordered Humphreys to throw back his left, and form a line oblique to and in rear of the one then held. This last attempt to save a piece of Sickle's advance line was attended with fearful consequences.... Humphreys at last received orders to fall back to the ridge between Cemetery Hill and Round Top. Barksdale, Wilcox and Perry followed, pouring in their fire with that of Alexander's batteries on Humphrey's front and flank. "I have never," says Humphreys, "been under a hotter fire of musketry and artillery combined. For the moment I thought the day was lost.... When Meade heard Sickles was disabled he ordered Hancock to take command of the Third corps in addition to his own. Hancock immediately led

Willard's Brigade of Hays' Division toward the left, and was about to advance to the support of Birney's Division when he found that it had all been driven to the rear (Author's Note: Bigelow's Battery, posted near the Trostle house, retired firing canister, which for a short time checked the advance of the enemy). At 7:30 Barksdale was wrapping his brigade around Humphrey's left. Willard charged and drove back Barksdale's right, but as he pushed on, the cross fire of Alexander's Artillery and Barksdale's retiring infantry was too severe. Willard's Brigade retired in good order, but with the loss of its commander and a large part of his brigade. At Hancock's request Meade had sent two divisions of the First corps, but they had not yet filled in the space between the left of Hancock and the right of Sykes.

This space was swept by the fire of the Federal artillery, which, when the infantry had been driven out of the Peach Orchard had been skillfully posted on the slope ... to cover their retreat. But it could not keep down the fire of Alexander's guns nor check the advance of Barksdale, who still pushed on around Humphreys left (What follows refers to the Twenty-first Mississippi.) Just as Barksdale's men were emerging in disorder from the bushes at the head of Plum Run. Hancock came riding by. The First Minnesota, of Harrow's Brigade, was the only regiment at hand. Hancock ordered it to charge. Eighty per cent of this gallant regiment fell, but Barksdale was driven back with heavy loss ... To drive back the hostile hands on Humphreys left, General Meade in person brought up two regiments of Lockwood's Brigade of the Twelfth corps, which drove the enemy from the field and almost to the Peach Orchard. The remnant of Humphreys two brigades joined in this advance, and recaptured Bigelow's guns, which had been captured by a regiment of Barksdale's Brigade. Barksdale was mortally wounded."

"As stated, when General Daniel E. Sickles was wounded – near the scene, and at the time, of the capture of Bigelow's Battery by the Twenty-first Mississippi – General Hancock was ordered to include the Third corps in his command. The condition in which he found it, and his efforts to check Barksdale's advance – after Wilcox had "gone astray" – are thus told in his official report: "General Birney informed me all of his division had been driven to the rear ... The number of General Humphreys's Division collected was very small, scarcely equal to an ordinary battalion. . . I established Willard's Brigade at the point through which Birney's Division and retired and fronting the enemy, who retired and fronting the enemy, who were vigorously pushing on. The brigade (Willard's) soon became engaged, losing its commander and many officers and men." The story of the Willard-Barksdale combat is thus quoted from the official report of the commander of the former: "The Rebels fired on the brigade as it fired on the brigade as it advanced, which fire was returned by a portion of the brigade as it advanced without halting. Many fell in the charge through the woods. Reaching the base of the hill the brigade continued to advance under the fire of a battery higher up on the (Peach Orchard) hill, and a concentric infantry fire on the right. The commander of the brigade finding it unable to stand so severe a fire, ordered his regiments to retire to its former position."

Hancock is quoted again, touching the fight between the First Minnesota and the Twenty-first Mississippi: "Riding along the line I met a regiment of the enemy, the head of whose column was about passing through an unprotected interval in our line ... The First Minnesota charged this regiment in handsome style, capturing its colors and driving back in disorder." This fiction of "captured colors and disorder" will do for a bulletin, but Captain Coates, commander of the regiment, knew better. His report reads: "To check the enemy we were ordered to advance, which we did, moving at double quick down the slope of the hill right upon the enemy. The fire encountered was terrible, and although we inflicted sever punishment upon the enemy and stopped his advance, we there lost in killed and wounded more than two-thirds of our men." In the list of officers named and killed and wounded, the Colonel, Lieutenant-Colonel, Major and Adjutant are listed with the latter. The brigade, except the Twenty-first. In no other was the "team work" so poor. From all the counts in the indictment of ineptitude and error. Barksdale's Brigade is free - it neither failed nor faltered, boggled nor wobbled. It alone did all that was required of it, and more. It could have rested on the laurels of beating more than double its force, in fair and open fight - with carrying the Peach Orchard position, the most conspicuous and brilliant success of the whole battle. But after this, it joined with Anderson's Brigades and drove the division of Humphreys from the field. And after that, three of its regiments met and bested Willard's Brigade of fresh troops, of Hancock's Corps. Its other regiments at the same time, was capturing two batteries and repulsing the First Minnesota of the corps, finally retiring, fighting, before Lockwood's Brigade of the Twelfth Corps, led by General Meade.

The assertion is made that no other brigade, in that battle, or any other, has been made such a record two hours of fighting - of assaulting three separate positions, engaging troops of three corps, four divisions, of Ward's, De Trobriand's, Graham's, Burling's, Brewster's, Carr's, Willard's, Harrow's and Lockwood's Brigades; and turning its back on none.

"For the moment," said General Humphreys, speaking of the time before Willard came up. "I thought the day was lost." "About 7:15," said General Hunt, speaking of the crossing of Plum Run and capture of Watson's Battery by Colonel Humphreys with the Twenty-first Mississippi, "the field was in a critical condition." But that "Anderson's Brigades "Went astray," what they apprehended would have been – the hole that Barksdale made in the enemy's line would never have been closed. Meade's retirement would have inevitably followed – the 3d of July, that dire day of the Confederacy, would not have been closed in disastrous defeat. But it would not be -Gettysburg had been written in the book of fate, as we read it. But the glory that Mississippians achieved on Peach Orchard Hill remains undimmed. And when the State erects a monument on the memorable field it will be placed there surmounted by the effigy of William Barksdale.

J. O. M.

Vicksburg, Miss.

* * *

CONFEDERATE VETERAN
JUNE 1911, 286
SOUTHERN HISTORICAL SOCIETY PAPERS
VOL. 38, 312-18

PVT. WILLIAM YOUNGBLOOD, 15TH ALABAMA VOL. INF.

[Ed. Note: Youngblood was a courier for Gen. Longstreet on July 2, 1863 and one of only two witnesses to indicate Lee's presence at the Sherfy farm.]

Personal Observations At Gettysburg
Unwritten History of the Gettysburg Campaign

By about 2 p.m. General William Barksdale's Mississippians had taken position. I was sitting on my horse within hearing distance of Generals Lee, Longstreet, and Hood. General Lee was standing on the ground and an orderly was holding the reins of his horse, while the others were in their saddles. General Hood said to General Lee, "my scouts report to me that there is a wagon road around Round Top at its foot over which I can move my troops. I believe that I can take one of my brigades and go around this mountain and attack them from the rear and capture it." General Lee asked General Longstreet's opinion. Longstreet said "I have great faith in General Hood's opinions and his ability to do whatever he plans to do." This was all the reply that Longstreet made. General Lee said, "Gentlemen, I cannot risk the loss of a brigade. Our men are in fine spirit and with great confidence will go into this battle. I believe we can win with a direct attack." Extending his hand to General Hood he said "God bless you General

Hood drive them away from you and take Round Top and the day is ours, and with tears in his eyes he turned, mounted the iron gray and rode away. Longstreet dispatched his staff and couriers along the battle lines to watch movements and report to him. He ordered me to remain with him. Thirty minutes after General Lee left us the cannons gave the signal to attack.

General Longstreet and I started out and found that we were in front of Wofford's men. I called General Longstreet's attention to this, and suggested the danger of being shot down by our own troops. He checked his horse and held him until Wofford's men had gotten in front of us. The Peach Orchard was on Wofford's left and Barksdale's right. General Longstreet, from the moment he came into the open field had his glasses constantly on Round Top. He was probably very concerned about a flanking attack as Barksdale's troops were moving on a leftward slant, leaving their right very unprotected.

Upon approaching the orchard the Union forces had fallen back. Our people were driving them but General William Barksdale's brigade had halted behind the small breastworks that the enemy had constructed. I called General Longstreet's attention to it and asked, "Do you want General Barksdale to halt?" He looked and said, "no, go and tell him to retake his position in line." I turned my horse and dashed to General Barksdale. I found General Barksdale on his horse standing in back of a brick milkhouse, and giving him the order from General Longstreet he put spurs to his horse, dashed a little ways along his line, giving the order to charge at double quick, when I distinctly heard a shot strike him and saw him fall from his horse. I went back to General Longstreet, who was guiding his horse into the peach orchard, told him of Barksdale's fall; when he said, "Go on beyond this orchard and tell General Alexander to advance his artillery, and to keep in touch with Wofford's left....

No state ever furnished braver nor better soldiers than that grand old State of Mississippi. No troops were ever commanded by a braver man than General Barksdale. Wofford's, Kershaw's, and Law's Brigades were beyond reproach, as game and true as ever carried a sword or gun.

* * *

EXCERPT FROM:
E.P. ALEXANDER, FIGHTING FOR THE CONFEDERACY,
GARY W. GALLAGHER, ED., 1989, 238-40

Col. Edward Porter Alexander
Commanding Longstreet's Artillery July 2, 1863

Now the weakest part of Sickle's line was the angle at the Peach Orchard. And it was the nearest, & time was of great importance, for it was late in the afternoon. I think the best & strongest assault would have been for both divisions, in a decidedly shorter line, to have simultaneously assaulted the Peach Orchard. But, as it was, Hood moved first and alone, & had a long & desperate fight upon Sickles's left flank, before McLaws was launched upon the Peach Orchard. I had hoped, with my 54 guns & close range , to make it short, sharp, & decisive. At close range there was less inequality in our guns, & especially in our ammunition, & I thought that if ever I could overwhelm & crush them I would do it now. But they really surprised me, both with the number of guns they developed, & the way they stuck to them. I don't think there was ever in our war a hotter, harder, sharper artillery afternoon than this. [Ed. Note: Gilbert's and Moody's batteries were placed directly in front of Barksdale's Brigade by order of Gen. Longstreet over McLaws' objections]

Gilbert's (Rhett's) battery of four guns had two fairly struck by the enemy's shot & dismounted. Of less than 75 men in action he had 40 killed & wounded, an unusual proportion of the wounds too being noted by surgeon's report as severe.... Within a half hour after beginning, Capt. Moody reported that he was short of force to handle his 24 pr. howitzers on the rocky ground & asked permission to try & borrow some volunteers from Barksdale's Miss. brigade lying close in our rear. I went to Gen. Barksdale & we soon got 8 fine fellows. Five of the 8 that night were dead or severely wounded. And at last seeing that the enemy was in greater force than I expected, I sent for my 8 reserve rifles, to put in the last ounce I could muster. While they were coming McLaws's infantry at last charged the Peach Orchard, & the enemy abandoned it & fell back.

When I saw their line broken & in retreat, I thought the battle was ours. Of course, I had known it was going to be all along, but now the hard part of it was over. All the rest would only be fun, pursuing the fugitives & making targets of them. I rode along my guns, urging the men to limber to the front as rapidly as possible, telling them we would "finish the whole war this afternoon."

* * *

DUBLIN (TX) PROGRESS,
AUGUST 8, 1913.

C. C. CUMMINGS
17TH MISSISSIPPI INFANTRY

Gettysburg

(Mr. C.C. Cummings in the *Fort Worth Record*. Mr. Cummings has addressed the Confederate Veterans of Dublin.)

That night of the second, there was gathered in a tent in a field hospital on the grounds all the field and staff of our regiment, the seventeenth Mississippi, except Dick Jones, the adjutant killed on the field. Colonel Holder was there with a wound in the groin; Fizer with a wound in the leg, as he preferred, and the writer, sergeant major, in the hand, as I preferred; Major Pulliam also in the hand, and Brown Jones, the Colonel's orderly, about fifteen in the body. We bivouacked at Willoughby Run on July 1st, where was opened the first day's battle and where a monument has been erected to the memory of the Federal General Reynolds, killed there that day. I claim to be the original reunionist, of the Gray and Blue. For when I was shot in the hand and while the blood was filling my buck gauntlet (I put on my best to die like a soldier,) two New York Zouaves running my way to avoid capture by our gray boys, who had broken through a barn on my left, and who would have served them roughly for the resistance they made. I beckoned them to come to me and I flung my arm around the neck of each, making a wounded squad, which our long experience told us would save us from being fired upon, and so I came off the field with my two prisoners. Wouldn't it be passing strange if I should meet one or both after fifty years at Gettysburg? Sickles will be there and they were Sickles men. Sickles and I exchanged compliments that day. I gave him my hand and pulled his leg. All the third day of July, 1863, we lay prone of our wounds in the hospital tent on the field of Gettysburg and heard the thunder of this legions or guns – big guns – from both sides, such guns as were never heard on the American continent. It was what has gone into history as "Pickett's charge" and while there were others, who crossed the deadline with Pickett to the high water mark, yet old Virginia our dear old mother, bore on her sacred bosom the bosom of war for us all. So let her son Pickett go down in history as the hero of us all.

* * *

Fort Worth Daily-Gazette,
May 31, 1887.

C. C. Cummings
17th Mississippi Infantry

Evening Services

The evening services were held in the opera house which, in spite of the intense heat, was filled to overflowing by the youth and beauty of our city who came to grace the occasion and by the elderly who came to join in rendering homage to the memory of the departed dead. The stage was becomingly arranged. A large cenotaph stood near the footlights over and around which was folded an American flag. This was in memory of the unknown dead who sleep in unmarked graves. A little back and to the right hung the picture of President Lincoln; beneath and to the right was a picture of General Grant, and underneath it a picture of Garfield. Directly underneath was a picture of President Arthur and to the right was that of Lee and a large steel engraving of Lee and Jackson. In the center of these pictures was a map showing the position of the troops at the Battle of Gettysburg on July 3, 1863, at 3:10 p. m. The map was planned by Judge Cummings in a few preliminary remarks. Above the map was the picture of two soldiers, one in blue and one in grey, lying, as thousands have lain, cold and dead upon the battlefield. The map was of great assistance to the audience enabling them to fully appreciate the situation. In front of the stage the Post was seated: Confederate veterans were seated at the right front of the stage; Texas and Mexican War veterans were assigned the seats of honor in the boxes; Commander Emerson, Adjutant Brown, the representatives of the press, the vocal music corps Elder Gavin and his reverend gentlemen companions, His Honor the Mayor, officers of the city government and members of the City Council, the Hon. Judge Cummings, and a volunteer platoon composed of Texas, Mexican, National and Confederate war veterans, were seated on the stage.

The orator of the occasion, Judge Cummings, in a few preliminary remarks explained the map of the Battle of Gettysburg. There was considerable of the grim irony of fate in the Judge's remarks as he pointed out where the various generals and battalions were stationed on that eventful day. His empty sleeve hanging by his side, a relic of that dreadful conflict, gave silent evidence that the Judge knew what he was speaking about. It was a graceful act in the post to select one who had opposed them on that fateful day, and a wise act in the post to select a gentleman whose oration will add much that is splendid and valuable to the war literature of the country. Judge Cummings said:

Soldiers of the Grand Army of the Republic:

Let us glance once more at these lines when this charge was ordered. As senior sergeant of the Seventeenth regiment, Barksdale's brigade of Mississippians, I by order numbered the guns just before we started in, 417, in line. In three hours 275 went down in the storm. My little company B, had twenty-one, out of an original 100 more, not a man over twenty-five, but one married. By 7 o'clock seven were dead, seven wounded, two to death, and I and seven were alone of those left to tell how they died. They died, as in their boyish grandiloquence in camp and march they ever said they would thus.

"*Whether the scaffold higher in battle's van.*

The littlest place for man to die is where he dies a man."

When out in the wheat field you shattered our batteries with your shot and shell. These brave boys marched forward to support them and fell like grains before the scytheman's stroke. Volunteers must come and man again the wrecks of battered batteries. Then steadily they rose and went forward to certain death. Calm as a May morning were they in their "last full measure of devotion."

* * *

EXCERPT FROM:
PUBLICATIONS OF THE MISSISSIPPI HISTORICAL SOCIETY
VOL. XIV, 1914, P. 238
J.S. MCNEILLY, "BARKSDALE'S MISSISSIPPI BRIGADE AT GETTYSBURG,"

MAJOR J.B. GERALD
18TH MISSISSIPPI
WACO HERALD

My recollection is that on the evening of the second day's fight we turned to the right from the road on which we had been marching into the woods and after moving through the timber for some distance halted and the order was given to "strip for the fight." The men carried their scanty change of clothing wrapped in their blankets and thrown over their shoulders; each regiment piled these in a heap and each left a man with the baggage. The field officers dismounted from their horses, the reason for this being that an order had been issued some time before that no officer below the rank of brigadier or acting brigadier-general should ride into battle, because of the fact that the government had a great deal of difficulty in replacing the horses killed. I gave my horse and watch as well as some other belongings to my servant. After these orders had been complied with the order was given "Dress to the colors and forward to the foe!"

Major George Gerald of the 18th Mississippi Infantry

(Baylor University - Texas Collection)

After moving through the woods a short distance, we came to a fence around a field of grain; the battle was progressing.

Before us lay open fields dotted with houses and right in our front were some farm houses with a grove of trees to the left, and the enemy drawn up in a double line of battle some five or six hundred yards distance and supported by artillery. We steadily advanced, driving the enemy before us until we reached the houses with the trees on our left, the trees proved to be a peach orchard. On the end of the orchard was a barn in which a part of the enemy had taken refuge. I was on the left of the regiment and the colonel and lieutenant colonel were somewhat to my right and partially protected by other buildings, and I with most of the regiment was directly in front of the barn. . . . . I called to the men that the barn must be captured and to follow me and I would open the door. They followed me with a rush and I forced the door open, and within less than two minutes we had killed, wounded or captured every man in the barn. The barn was filled with smoke so dense that it was nearly impossible to distinguish a man's body in it, such a continuous fire had the enemy within kept up. We left the barn and the brigade moved through the orchard towards the heights, still driving the enemy before them. General Barksdale encouraged the men by shouting, "Forward, men, forward," which was the only command that I ever heard him give after a battle commenced. By this time we were under heavy fire of two lines of battle and their artillery and our losses had been very heavy, and recognizing the impossibility of breaking these fresh lines of battle, we fell back to the orchard, where we spent the night. We retreated to the orchard in fairly good order. It was nearly dark when we went into camp at the orchard.

* * *

DAILY NEWS (MERIDIAN, MS)
APRIL 13, 1901.

PRIVATE JOSEPH C. LLOYD
13TH MISSISSIPPI INFANTRY

None braver from any State than Gen. Wm. Barksdale. I was a participant in that Gettysburg charge – a member of his brigade – was wounded there a few moments after Barksdale was, and the last man that saw him on the field.

There is no battle, no event in the history of the Confederacy, more indelibly stamped upon my memory that this battle, this particular charge at Gettysburg.

Every reference by Mr. Maury to the different states, positions, maneuvers, &c., during the charge brings the picture back to my mind, as fresh as the day it occurred.

That rocky ridge where we laid in line of battle for quite a while and from which we started – that beautiful oak grove just behind us.

Private Joseph C. Lloyd of the 13th Mississippi Infantry

(Children of John Emory Lloyd and Olive Sparks)

The Washington Artillery just in our front. That terrific cannonade that followed the challenge of the Washington Artillery that lasted far more than an hour.

The crash, crash, among the trees and the falling of the branches.

The terrible, trying ordeal of a line of battle lying down waiting for the more terrible ordeal to come "forward."

The last close examination and scrutiny of the field by the Generals. Their comments, suggestions, counsel, &c. Longstreet, Barksdale and even down to our brave idolized Colonel Carter of the 13th Mississippi.

Their last drink, in which Carter did not join, before giving the order "forward" – all occurred just in front of our colors.

"General, move your division forward," came from Longstreet to McLaws, followed by a suggestive grasp of hands, "Attention" brought us to our feet and "Forward, march!" followed at once.

I shrink at the recollection now and hope – our sons – will never have to undergo such an ordeal.

Our receiving the first volley from the enemy, the ringing order came, "charge bayonets!"

With that electric shout, that only the Confederate troops could only give, the "shock of battle" came in its fullest meaning.

The three hundred yards in front of the Turn Pike referred to, was made in short order.

At the Turn Pike we found two rows of rail fence, ten rails high, as I recollect, the Zouaves referred to, behind them.

Some of us scaled the fences and some tore them down. Both fences and Zouaves disappeared quicker than I could possibly give an account of it. Oh that Turn Pike, that death trap, a perfect torrent of death racing down it.

Our losses were very heavy there. Our Colonel Carter was killed in it while urging his men not to stop in it, but to cross over into the field beyond. While charging through that field, we caught the cross fire referred to by Maury, both from artillery and musketry, and in the "right wheel" assisted in the capture of the battery. The left of our regiment also captured 200 prisoners in the peach orchard and about the houses, as stated.

It is likely that we advanced faster than the troops on our right, but the course of the ravine running diagonal to our general line of advance, would naturally cause troops on our right to reach the ravine before we did and meet another line of the enemy. The passage of that ravine was to be retreated along its whole length.

We had to charge further but when we did reach it we found the resistance. A fresh line of the enemy were in waiting for us and did charge us.

Here is where our first wavering began and where General Barksdale appeared right among us. A part of our men had even entered the thick undergrowth, when their charges down the hill on the opposite side was made.

Barksdale's appearance among us begins my sequel. We were just making an effort to reform, our line was very thin, in fact it could not be called a line at all.

He was on his horse with drawn sword. "Rally men! Rally! No retreating! No running! &c." The fresh line had charged down into the bushes and were firing on us, and with this trouble on our hands there was no time to rally, and we began returning their fire as best we could, I do not remember seeing any efforts among us just then, though we needed no orders. It was "fighting" and on mighty close quarters. Pretty quick word passed "Barksdale is shot." As I turned to look, his horse wheeled and showed his back and I could see where the ball came out at the point of the shoulders. Being about to fall, he was taken off his horse, and several started, and I thought, did carry him from the field.

I turned my attention again to the bushes, but it must have been that nearly all of the few of us there had retreated with Barksdale, and I noticed only a man occasionally.

There are times when a proposition can be debated, but it took me only a second to decide that a retreat was best for me, and I was going to act on it.

In turning for a last look at the bushes, I discovered their line of battle, coming out in thirty feet of me, the question seemed debatable at least.

I saw but one man anywhere near me just then, and he was running for all he was worth. I saw one Yankee shoot at him, but he kept running. Another one raised to shoot at me, and at the crack of his gun, I know that he had not missed. The touch of that "minnie" closed all debate, and I sat down and threw up my hand, surrendered. They advanced over me and I immediately went down into the bushes, for a comfortable place. To be away from further danger, and perhaps to die – I was afraid I might bleed to death.

I remained there a half hour or more, when their line retreated back over me.

One soldier turned back to me and proffered assistance and kindly took my handkerchief and made a sling for my arm, then bade me "goodbye." Believing that no Yankees were between me and our lines. About 50 yards. I suppose, from the bushes, I heard a call and turning to it I found it to be Gen. Barksdale. He was reclining and asked for water, when I turned my canteen bottom up to his lips, and to my surprise there was not a drop in it. I found a bullet hole in it which accounted for it.

He was perfectly conscious and gave me messages to his Brigade. Said he was mortally wounded and would die very soon the death of a true and brave soldier.

I gave him my hand and parting and assured him that I would direct litter bearers where to find him as fast as I could. He was entirely alone and I am fully satisfied that I was the last man from our side that saw him on the field.

I found later in making my way out that I was not outside of the Yankee lines. Ran right into one of their lines but it being dusk I was not noticed. I tacked back and tried another direction and finally got to our field hospital about 10 o'clock, two miles in the rear.

Gen. Barksdale was taken to Gettysburg and died the next day. J.C.L.

P. S. – I would love to write of Barksdale at great length. His sociability with his men in camp, his sympathy for them on the march, his impetuous bravery on the field of battle, all of it acted as an inspiration to his men.

I never had but one objection to him. It always seemed to me that he was too eager, too anxious to get into a fight. J. C. L.

* * *

CONFEDERATE VETERAN
NOVEMBER 1901

MAJ. JOHN J. HOOD

In an address to the Barksdale Camp of Sons of Confederate Veterans at Meridian, Miss., paid tribute to the gallant Barksdale while in command of his Mississippi Brigade at Gettysburg:

Emerging from the peach orchard into the open plateau beyond, he [Ed. Note: Colonel B.G.Humphreys] discovered that the brigades to Barksdale's right had been held in check by the enemy. Off to the right some three hundred yards another battery was thrown into position to enfilade his column. In a flash "Right wheel, charge!" rang above the roar of battle, and with the rattle of muskets, the clang of bayonets, and the shouts of victory Winslow's Massachusetts Battery fell a prize to the sons of Mississippi. Still the battle raged. The Federal general, Graham, retreated across Plum Run, and rallied his line behind a battery one thousand feet away beyond the ravine. If that battery can be silenced and Graham routed again, the enemy's line will be completely severed. If severed, victory is ours. Mayhap in this charge hangs the history of the Confederacy. What a moment! The command is given and the regiment [Ed. Note: 21st MS] moves forward into the ravine, and under its shelter reforms in line of battle. Up the other side, and out upon the plateau at the base of the ridge, they advance for the last and most desperate charge.

Sickles rode to the front to encourage his men. Up the slope the Mississippians advanced, undaunted by bursting shells or screaming Minies. Thinned by the storm which swept down with such terrific fury from the ridge, the advance line staggered and began to waver. The awful crisis of that awful day had come. As on a thread hung the hopes of s struggling nation. Amid the shifting scenes of this doubtful battle the very course of history swung to and fro in the trembling balance. With its ebb and flow rushed the tide of our country's hopes.

Suddenly far to the left, a wild shout of victory rings above the din of battle. Looking back toward the turnpike, the majestic form of the heroic Barksdale, like the fabled god of battle, towering in the front of the advancing column, was seen leading the charge on the last line of the enemy at the ravine. The inspiration spread like contagion through the regiment, and with a yell that sounded to the very dome of heaven they rushed upon the enemy's line. As the autumn leaves are scattered by the blasts of winter, so was the Federal line swept away by the charge of that hurricane. On, right on, until the fiery breath of the cannon was hot in their faces they pressed. The brave gunners who manned that battery never deserted their posts, and died with their guns beside them. Sickles fell desperately wounded, but escaped into his line. Graham, who had stood by his men till the last had fallen, fell into the hands of the Mississippians, a prisoner of war. The last line was carried, the last battery was captured, the enemy's line was cut in twain.

Thus it was that the Twenty-First Mississippi Regiment bore the stars and bars to the very farthest point reached in the enemy's line on the bloody field of Gettysburg. Holder and Griffin had swept across the turnpike in the pursuit of the Red Zouaves, and were now making the last desperate assault upon the enemy's line. Barksdale stood on the turnpike watching the movements of his heroes as they drove the enemy before them. His bright eyes flashed as though lit by a spark from the fires eternal. His face beamed with the glow of glorious victory. His thin lip curled with that haughty smile which meant defiance. He stood the perfect picture of a true hero. Far to the right he had watched the gallant dash of the Twenty-First Regiment, and saw the enemy's line broken and scattered. He saw its lines re-forming in his front across the ravine, and drawing his sword, he sprang to the front, the flush of victory on his face, shouting: "They are whipped. We will drive them beyond the Susquehanna." Then it was the cheer went up which had lent the inspiration to Humphreys's wavering column. They charged the ravine as Mississippians were wont to charge, but the Federal lines, reenforced, had rallied beneath its shelter, and poured volley after volley into his lines. His column, halting beneath the deadly fire, reeled to and fro like a forest beneath a tempest. The gallant Holder fell desperately wounded. Griffin fell. Fizer fell. Once more they dashed against the enemy. Then, waving his sword in the very thickest of the fray, in the last charge upon the enemy's last line,

Where swords were clanging, clashing,
Where balls were crushing, crashing,

This soldier, patriot, hero fell. Amid the wild roar of the cannon, the fierce scream of shrapnel, and clang and glitter of sabers, midst all the pomp and glorious circumstances of war --- the great soul, the unconquered spirit of the immortal Barksdale winged its flight. No shaft marks the spot where he fell. The Federal authorities refused to allow the point they reached to be designated by appropriate stones, but that gallant charge is written upon the hearts of his countrymen, and will be told in song and story as long as gallant deeds and heroism are virtues.

* * *

HOUSTON POST,
JANUARY 14, 1900.

SERGEANT J. T. CONE
13TH MISSISSIPPI INFANTRY

Barksdale's Last Order

One by one the old Confederates disappear and soon will all be hushed in death, and then our children will say: I've heard my father speak of such and such a fight and what a hard time he had in prison, etc."

It was my luck to be in Lee's army and the last battle I was in made the deepest impression - that was the battle of Gettysburg. Our command got into it the worst - Texas brigade on the right, the Mississippi brigade next and the Alabama next. These brigades had a hard time that evening. I belonged to the 13th regiment and it was our duty to dislodge the bluecoats from a large barn. As we went into the fight Nute [Newton] Nash turned to me and said, "You know who is in the habit of running, keep them in line."

I'll say it was just about all I could to keep myself in line and let them be. Poor Nute! That was the last word I ever heard him utter. He was killed. Clem Black, Craig Davis, Hock Brister and others of our company fell that evening and many were wounded. John Durham our color bearer. He lost an arm and had to heave the colors on the field. Jack Boyd bravely picked them up and carried them to the front. Last, but by no means least, our brigadier general, William Barksdale, fell. Much good blood was spilt that evening, but none braver than General Barksdale (he was the first year of the war our colonel.) I suppose I was the last man who ever heard General Barksdale give orders. His last word was "Forward." I looked around to see where our line was and we had none. The blue coats came with a solid line. Barksdale was killed, poor me. I just had to obey orders and forward to Fort Delaware to stay until June, 1865. J. T. Cone. Company I, Thirteen Mississippi Regiment.

* * *

HOUSTON POST,
JUNE 24, 1913.

PRIVATE D. G. MAGGARD
13TH MISSISSIPPI INFANTRY

One of the Veterans of the Battle of Gettysburg

D. G. Maggard of 1311 Sixth Street is one of the few survivors of the Battle of Gettysburg. He joined the Confederate army when a beardless youth of 17. So determined was he to fight for the cause of the South that he ran away from the private school he was attending in Northern Mississippi and enlisted against the wishes of his father, who wanted him to wait until he arrived at the age of 18. The age at which young men were conscripted. He became a member of Company B, Thirteenth Mississippi Regiment, Longstreet's corps of the Army of Northern Virginia.

I first saw active service on the second battle of Fredericksburg, says Mr. Maggard. We had taken a position at Marie's Heights close to Fredericksburg. A large detachment of Federal forces crossed the river on pontoon bridges, and seeing that we would be cut off from the main division we were forced to retreat to the top of Marie's Heights, firing as we went. In this battle a thick army blanket I had strapped around me saved my life. The bullet passed through a letter I had received from my sweetheart, but did not quite reach my flesh. I still have the bullet and the letter.

After we had driven the Yankees out of Fredericksburg we had a long, hard march to Gettysburg arriving there on the night of the first

Private D. G. Maggard of the 13th Mississippi Infantry

(Houston Post)

day's fighting. We were thrown into line of battle on the following morning and remained in a lying position until 3 o'clock in the afternoon when the order was given to make the memorable charge. We went some distance before seeing the enemy. The first we encountered were a detachment of New York Zouaves, costumed in their red bloomer like trousers and decorated coats. Several of my comrades were killed in this early skirmish, but we soon compelled the soldiers from New York to retreat from behind the plank fence where they had taken their position. One of them, seeing that he had little chance of escape, crawled into a big haystack. I saw his protruding feet and yanked him out. I ordered him to the rear, and the last I saw of him he was going up the hill we had just descended from.

We soon found ourselves in the thick of battle. I saw both Northern and Southern soldiers falling around me in great numbers. The last I saw of General Barksdale he was leaning against his horse mortally wounded. He was game to the last, and I remember him calling out, 'Boys, I am a dead man, but charge 'em, damn 'em, charge 'em, and don't fall back!' Colonel Carter was also killed in the charge. I recall seeing General Longstreet riding along the lines on his beautiful horse urging his men on and issuing orders. He was a grand specimen of manhood, and I saw him that day with his plumed hat and gray uniform riding his beautiful horse, I thought him the finest-looking man I had ever seen.

Those of our company followed our dying general, as we charged ahead the fight became fiercer and fiercer. Soldiers lay dying on all sides, and we were forced to trample upon scores. Our company was soon cut off from the others, and my comrades were falling fast. Before long I discovered myself surrounded by Northern soldiers and was disarmed. It is the wonder of my life to this day how I escaped uninjured. I was taken with many other Confederates who were captured to a 'bull-pen,' where we were kept under guard. The next morning we were forced to march to Westminster. I was barefooted and before long I was so footsore that I could hardly walk. It sure made me 'hot' to have a big Yankee behind me on a horse prodding me on with his bayonet when my feet were nothing but blisters. From Westminster we were taken to Baltimore Md. There were many Southern sympathizers in the city, and we saw Confederate flags unfurled from many of the buildings and houses. Many pretty girls waved their handkerchiefs at us that day and blew kisses to us as we marched by, prisoners of war.

From Baltimore we took a boat to Fort Delaware, where I remained a prisoner for 23 months and 17 days. They tried to induce me to take the oath of allegiance, but I told them with spirit that I would stay there 'until my hair was white as cotton before taking their darn oath.' General Schofield, who was in charge of the island, upon hearing me say this, said, 'that's the darndest little rebel I ever saw.' He took an instant liking to me and placed me in charge of one of the wards in the hospital. It sure did me good to be able to 'boss' some Yankees around. I received very good treatment while a prisoner of war at Fort Delaware. In May, 1865, I was paroled and sent home.

Mr. Maggard has been connected with the Houston Transfer and Carriage Company for a number of years. He is at present superintendent of buses for the company, and is also in charge of the Bureau of Information at the Grand Central Station.

* * *

NEWTON (MS) HERALD,
AUGUST 4, 1927.

PRIVATE T. M. SCANLAN
13TH MISSISSIPPI INFANTRY

The next great fight my father was in was Gettysburg, where he was wounded in the second day's fighting, as Barksdale's brigade participated in the famous, "Peach Orchard" Charge. He lay on the battle field for thirty-six hours. Right here happened one of those thrilling things that mark the devotion between comrades during wartimes. My father was badly sot just below the right knee. The confederate surgeons came along and examined his wound. The condition in which they found it, required a certain surgical instrument which they did not have. A group of Federal doctors were at work nearby and the Confederate surgeons borrowed the necessary instrument from the Federal surgeon. The doctors then began to get ready to cut off my father's leg. He protested against the procedure and just at the moment a member of my father's Company, named Moore, from Newton County, came along, engaged in litter-bearing service. My father appealed to Moore, begging him not to let the doctors cut off his leg. In the face of the protests of the two men the doctors agreed that they would probe for the bullet and determine the extent of the wound. Again they insisted that the limb should come off, whereupon my father's friend, drew his old horse pistol and boldly asserted that he would shoot any man that stuck a knife in the limb. The doctors indignantly left my father and he lay on the battle-field for quite a while, when he was picked up by the enemy and carried to the field hospital in Gettysburg. Here might be another interesting occurrence: My father had an Aunt in the State of Ohio, who had two sons in the Federal army. He wrote her of his condition and asked her to come to Gettysburg. She responded immediately and no doubt it was owing to her kind ministrations that my father fared as well as he did. This Aunt was a woman of great executive ability. She invoked the aid of some Federal Major-General, who procured for her a small company of soldiers to help her look after the wounded, both friend and foe. She appealed to the women of Gettysburg for their assistance and their contributions of towels, sheets and linins, or anything else that would make bandages, and those noble women went forth into the hospitals and whenever the wounded might be found, and ministered unto them, no matter what side they fought.

My father finally became convalescent and was carried to Governor's Island as a prisoner, and was finally exchanged and sent to Richmond, Va., He was permitted prematurely to be exchanged. In fact a Commandant of the force refused to let him go, asserting that he was not physically able to make the trip; whereupon the Commandant's little daughter, a child seven years old, clasped her arms around her father's neck and pled with him to let the young rebel prisoner, even though so badly wounded, make the trip to Richmond. Under the pressure of the child's appeal the Commandant permitted my father to go onboard the ship that was carrying to Richmond the prisoners for exchange.

* * *

EXCERPT FROM:
PUBLICATIONS OF THE MISSISSIPPI HISTORICAL SOCIETY
VOL. XIV, 1914
"BARKSDALE'S MISSISSIPPI BRIGADE AT GETTYSBURG,"
J.S. MCNEILLY

COLONEL WILLIAM D. HOLDER
17TH MISSISSIPPI INFANTRY

Gettysburg – Letter From Colonel Holder.

In commenting upon a Gettysburg incident in a recent issue, we inadvertently stated that the 17th Mississippi Regiment was commanded by Lieutenant Colonel Fizher; thinking that Colonel W. D. Holder, its distinguished colonel, had been wounded in the previous battle of Marye's Hill. Immediately after publishing, and on a moments reflection, we remembered that he was wounded at Gettysburg. The following is a reply to a letter correcting this error. We gladly gave it publicity; as it would be a grievous wrong to deny to any of the regimental commanders of Barksdale's brigade on that memorable occasion, such renown as they won there:

JACKSON, MISS, SEPT. 18, 1886
CAPT J. S. MCNEILLY, GREENVILLE, MISS.:

My Dear Captain – I thank you for your esteemed favor of the 14th inst., enclosing article in reference to the battle of Gettysburg. Replying, will say, that I had the honor to command the 17th Mississippi Regiment in the memorable charge on 2d July, 1863, receiving a wound in the abdomen, from which I was permanently disabled. Barksdale's brigade, in line of battle on the day in question, stood thus: Extreme left, 13th Mississippi Regiment, Colonel Carter; 17th, Colonel Holder; 18th, Colonel Griffin; 21st General Humphreys. Colonel Griffin was wounded in the leg, Colonel Carter killed on the field, and Gen. Barksdale fell in the midst of his gallant old 13th Regiment. He always gravitated to this beloved regiment in every battle. I regret that I cannot name the color bearer, who signalized himself at the "Trostle House." No color bearer of that gallant brigade, however, could be other than a hero. The prerequisites were "every inch a solder," and equal to emergency. They dared to you where duty called, and knew by the crucial test of many to an ensanguined field, that their regiments would stand by them to the bitter end. I am yours truly, W. D. Holder.

* * *

MCKINNEY (TX) DEMOCRAT
MAY 23, 1895.

PRIVATE WILLIAM M. ABERNATHY
17TH MISSISSIPPI INFANTRY

Colonel W. M. Abernathy

Hon. W. M. Abernathy of McKinney, colonel and inspector general Northeastern division of Texas, was private, company B, Seventeenth regiment of Mississippi infantry, Barksdale's brigade, McLaw's division, Longstreet's corps, Army of Northern Virginia, in 1862; participated in the campaigns of 1862, 1863, 1864 and 1865; was wounded three times and left because of wounds in a hospital at Gettysburg; carried to David's Island, New York, and exchanged as being too badly disabled to serve: rejoined his command and was detailed as courier at Longstreet's headquarters in the latter part of 1864 and surrendered at Appomattox.

* * *

McKinney Democrat
October 5, 1899.

Private William M. Abernathy
17th Mississippi Infantry

His Flag Now

An Interview With W. M. Abernathy
He Stands by the President and the Flag.

McKinney, Texas, September 30. – Ex-Confederate W. M. Abernathy, whose name was mentioned in the Fort Worth notes in The News, makes a statement, giving some facts regarding the Gettysburg battle. He also wants it understood plainly that he is loyal to our president and army. He says:

"I am glad to see Col Cummings has corrected the statement in The News that he and I belonged to the Seventeenth Massachusetts at the Battle of Gettysburg. Even here in my office I have hung up a record showing I was a member of the Seventeenth Mississippi and surrendered with Lee at Appomattox, we had a hard time explaining to J. Pearson Welch, Warden and some others the color of the regiment to which I belonged. Now, I guess I can go to the fair on ex-Confederate day.

"Yes, I was at Gettysburg, and the colonel has not told all. We were supporting Moody's battery and it lost so many men that volunteers were called for to help them man their guns. The colonel's cheery voice, 'Come along, boys,' was the first to answer. Lundy Gunn, a brawny sergeant of Company A, was second. Well before Cummings ceased speaking a dozen were running to the guns. If all good ex-Confederates get to heaven I'll beat the colonel there for he will have to rummage around the Pennsylvania fields for his hand before it starts.

"My old comrade is mistaken about fighting the Seventeenth Massachusetts that day. I walked over the ground and examined their monuments some two years ago. The first troops we encountered were from New Jersey. They were formed along the Emmitsburg Pike. Their monuments are in the peach orchard. Here we had another tough fight. Lieut. Ramrean was shot through the left arm and part of

Private William M. Abernathy of the 17th Mississippi Infantry

(John Hoopes)

his right hand shot off. He still cheered on the boys. After driving back the Indianans, I noticed him; he had been shot through the mouth. He was still in front waving that old black hat, blood oozing from his mouth. There were no well men helping the wounded that day. The wounded men went to the front and Jim R. was leading the left of the company in the charge on the Excelsior brigade and their batteries. They too, have monuments to commemorate the valor of the New York Excelsior's.

"About six hundred yards back of the New York Excelsior monuments is where our brigadier Barksdale, was killed. A marble tablet marks the spot. Back nearly the same distance the line of monuments to a Wisconsin command. Here they stopped our brigade to finish the job next day which they didn't do.

"I am coming to Dallas fair not to glory having helped drive those who carried the flag there. It's my flag now. I made peace at Appomattox. I am ready to jubilee when our boys carry it in triumph over in the Philippines. I am with our president. We will never lower it one inch or carry it back now one step."

* * *

McKinney Weekly Democrat-Gazette
January 31, 1907.

William M. Abernathy
17th Mississippi Infantry

Greetings From Old Servant

Hon. W. M. Abernathy sends the following from his bed of sickness.

"Quite a number of people asked me about my old servant Simon in Mississippi. I have had his letter several days and now let him answer:

"My Dear Friend – I received your letter and check on Christmas eve. I must thank you for remembering me. It does my heart good to hear from you as much or more than my presence. I am quite well and was glad to hear you and your family was the same. Well, you and I are one more year nearer our eternal home where I trust we shall meet and know each other. Please always kindly remember me to Mrs. A and the children. I am over always your faithful,

"Simon."

Who knows, old Simon, we are one year nearer the knowledge. Meantime, from my bed of sickness, I pass up rheumatic old Simon's greetings and good wishes to the *Courier-Gazette* and *Democrat Gazette*, their readers, our friends, and especially to that faithful comrade Joe T. Hannaford, Morrilton, Ark., whose name brings a smile and a God blessing. He sought me at Gettysburg and found me down but faithful Simon was already beside me, grieving piteously."

* * *

McKINNEY DAILY COURIER-GAZETTE
AUGUST 17, 1911

WILLIAM M. ABERNATHY
17TH MISSISSIPPI INFANTRY

W. M. Abernathy; An Appreciation.

Ft. Worth Lawyer Pays Fine Tribute – Tells Of Old War Times. Many Fights, Narrow Escapes, And Record Of Great Morality Among Their Company.

Ft. Worth, Tex., Aug 16, - To the *Courier Gazette*:

Some kind friend has sent me your paper of Aug 9, containing sketch of my old-time friend, neighbor and messmate in the Virginia army, Hon. W. M. (Billy) Abernathy, noting his sudden death on the 8th inst. I call him Billy, for such was he ever known to us of the Early Grove mess in Marshall County, Miss., where we enlisted in April 1861, as neighbor boys (not in 1862 as the sketch has it.) We were in it from the first and for half a century he has kept the faith as a true son of the South, than whom none braver or better under the flaming red cross to glory and victory. We stood together on that fatal 2nd of July, 1863, looking down into the valley of death in the Peach Orchard at Gettysburg from 8 o'clock in the morning till about 4 p. m., in the afternoon, on Seminary Ridge, awaiting the order to rush down on Sickles with his bloody angle of 10,000 men in blue with munitions of artillery and all the latest improved arms, which our single thin gray line was ordered to break and scatter, but at a cost of 275 of the 448 men in ranks of our regiment, 17th Mississippi Infantry. I was sergeant major of the regiment and numbered the guns going in, outside of the officers. In two hours these 275 were put out of action and when the night came there was gathered in a single tent on that field of glory all the field and staff of the regiment except Dick Jones, killed dead on the field - that is. Col. Holder wounded in the groin, Lieut. Col. Fizer wounded in the head, Sergt. Major Cummings, a similar wound in the hand and the orderly of the Col. Brown Jones, disabled also. When shot I was making for a battery in front which I failed to reach, but Billy heard the order to go for it and was among the captors of this death dealing machine of war and was shot off one of the cannons of the enemy after waving his hat in victory astride it.

Since Billy came to Texas, I have met him now and then at reunions and heard from him by letter often. In one of these he told me of the last days before the surrender how our company of originally 125 (B, 17th Miss,) was reduced to but seven at the wind-up, and that one day shortly before this, they went bathing and every one of the seven showed on their bodies honorable wounds received in defense of the constitutional liberty of our beloved Southland.

Billy was in every march and fight with his command from the first great battle of the war at Manassas (which the federals name more appropriately for them, Bull Run) fifty years ago last July, till the last shot at Appomattox, April 9, 1865.

He wrote me once of how these boys wept when they were told the war was at an end and without our gaining what we went out to battle for and that they smashed the guns against trees and stumps rather than suffer the disgrace of grounding arms. You may search the records of that great war and no company sizes up with a better showing for courage and fidelity to the cause than our little company of boys - Co. B, 17th Miss. Regiment, Barksdale's Brigade, McLaws' Division, Longstreet's Corps, Army of Northern Virginia, R. E. Lee, General.

Billy was well known to Marse Robert as one of his most trusted couriers and his children will certainly cherish this noble example of so worthy a sire as one so near to the greatest commander in all the annals of history for ability and morality and humanity.

Mississippi was the storm center of the great upheaval and these Mississippians will ever be proud of having acted so long and faithfully their part in maintaining local self-government and Home Rule, a cause that is never lost.

C. C. Cummings

* * *

CLARION-LEDGER
JUNE 12, 1896

COLONEL W. D. HOLDER
18TH MISSISSIPPI INFANTRY

Incident Of The War

How Col. Holder and W. L. McKee Fared At Gettysburg. Both Were Wounded and Went to the Rear, the Latter Holding His Horse by The Tail.

A few days ago Col. W. D. Holder was telling "the boys" about the State House of an instance of great self-sacrifice, wherein a young man

Colonel William Dunbar Holder
portrait by Andres Molinary

(Miss. Dept. of Archives and History)

---

from Madison county, a mere lad of 15 or 16 years of age, and a courier on the staff of Gen. Wm. Barksdale, had played the leading role.

Col. Holder told of how his horse had been shot from under him at Gettysburg, and how a beardless boy, seeing his predicament had galloped into the very thickest of the fight, and tendered his own wound. The offer was accepted and Col. Holder galloped along in front of the lines between the galling fires to try to prevail upon Gen. Barksdale to retire from the front as he was exposing himself unnecessarily. But Barksdale was killed and Holder severely wounded in the stomach.

Appreciating that he was severely wounded, Col. Holder made for the rear, and ran across the young man who had loaned him his horse. The boy was also shot in the foot, but he grabbed the horse by the tail and accompanied the Colonel to the hospital tent.

Having related their awful experience of the next few days, and trip to Winchester, which he lost sight of the brave lad Col. Holder remarked he would like to know what became of him.

"I can tell you," said Deputy Revenue Agent Alfred Galloway, "he was a neighbor of ours in Madison county and I've heard him tell that very incident."

Col. Holder wrote to the address furnished by Mr. Galloway, and a day or two since was made happy by the receipt of the following, and which he has been prevailed upon to have published for the information of other old soldiers: Abbott, Tex., June 7th, 1896.

Col. W. D. Holder, Jackson, Miss.

Dear Colonel and friend of long ago - I yesterday received a letter that seemed to me as a communication from the dead. For I had though for nearly thirty-three years that you were dead. Why? Because while you and I were stopping at Winchester, Va., at private houses, after our having been wounded at Gettysburg, and coming back to Virginia, - I was informed, and so credibly, that never for a moment doubted it, that your orderly - a young man whose name I have at last forgotten, but who I knew at the time was attending you, had, through mistake, in making a toddy for you, given you corrosive sublimate instead of whiskey, and that the poison proved fatal.

Your letter of the 4th inst. was the first time that I had ever heard of it, contradicted, nothing short of such directed information could have ever brought me to believe otherwise than that you had died at Winchester or above stated. Therefor imagine, if you can my glad surprise, on the receipt of your most welcome letter, a letter I expect to preserve as a sacred memento and souvenir.

Yes, Colonel, I am the identical boy to whom you refer, and when I read your letter, the whole picture presents itself afresh to my mental vision so perfectly, that I could give on the scene as vividly as if the events had transpired only yesterday - even in all its phases: the field of battle and carnage, the whizz of the minie balls, the hum and explosion of the shells; how the gallant Barksdale and his old tried and true Mississippi Brigade were driving and did drive three Yankee lines of battle from the field, capturing and passing over twenty-five pieces of Yankee artillery in their second march, starting into the fight only

1500 strong; how, when we only had struck the 4th line of Yankee troops, on account of depletion of our ranks after our gallant Barksdale had fallen between our lines and the enemies, the old tired and true brigade, the heroes of the laying of the Pontoon crossing at Fredericksburg, had to begin to fall back and retrace their steps across the field, they ad so lately crossed so victoriously. And amid it all how vividly is portrayed that little episode of ours, when you came up with me and said: "Billy, I am badly wounded, and I must go to the rear as quickly as I can." I told you "all right colonel; I am wounded, too, and I will go with you," and as there was no time for delay just at that time and place, I grabbed hold of my horse's tail, on which you were mounted, and made good my hold, until we got to a place hardly so much exposed, and when I could mount behind you, and our trip to the field hospital; also our Sunday night camp on the Virginia side of the Potomac at Williamsburg, my discovering the live things in my camp fire, and how it frightened me, and you laughing at me for my fright; our trip to Martinsburg, our stopping at the hospital there, and having the live things extracted from our wounds, and then on to Winchester.

There I thought you had died. I remained there two weeks, then went back to Richmond, was on my crutches about three months, finally got well enough to return to service, and rejoined my command at Chickamauga just in time to join the East Tennessee campaign.

I have had many a good, hearty laugh over how I came off the battlefield at Gettysburg. It must have been about the same day you wrote me that I was telling some gentlemen in town here about our trip from Gettysburg to Winchester, and about the worms getting in our wounds, and the circumstances, as I believed, of your death from poison at Winchester. How thankful I am to know that I have been mistaken all these long years, and that you are still living.

W. L. McKee
Company C, 18th Miss.

* * *

RICHMOND ENQUIRER
JULY 21, 1863

CAPTAIN ROBERT T. BROWN
18TH MISSISSIPPI INFANTRY

Captain Robert T. Brown of the 18th Mississippi regiment, reported as missing, after the battle of Gettysburg, was wounded in the left foot and left hand, and captured by the enemy. A letter from him has been received by flag of truce boat, giving this, and the further gratifying intelligence, that his wounds are not of a serious nature.

* * *

CLARION (MS) LEDGER
AUGUST 31, 1886

J. L. SMITH
118TH PENNSYLVANIA INFANTRY

Who Was the Color-Bearer?

Philadelphia, Aug. 31, 1886.

To the Adjutant-General of Mississippi:

Dear Sir: - At the battle of Gettysburg, July 2d, 5:30 p. m., Barksdale's brigade charged through the Trostle House yard, on the right of First brigade, first division, first corps. The flag of the leading regiment was borne so bravely by the color-bearer through the gate of the yard, and I saw him so plainly in advance of his regiment for that I am anxious to get the name of his regiment for insertion in the history of 118th Corn Exchange regiment, Pennsylvania volunteer now being written. If you can aid me in this you will greatly oblige. Yours,

J. L. Smith

118th Pennsylvania Infantry

Mississippi papers please notice.

Lexington Advertiser

We believe the color-bearer referred to in the above was Carrol Williams, of Company K., 18th Mississippi Regiment, who was killed soon after, the same day, and about the time Gen. Barksdale fell. We are not certain about the name of the house, but if it was the brick house where in a lot of Federals had taken refuge and were firing out of it on our men, and where Moody's Company of Artillery, which made the charge with us, unlimbered and were about to open fire on it, when a white flag soon indicated their surrender; and if there was a barn close by in which we took a good many prisoners, was the Trostle House, then Carroll Williams was the man who carried the flag at that particular time. Soon after, when Carroll received the almost instantly fatal shot, he turned to A. J. Guice who was close by and handed him the colors, and said to him, "take it and go on!" and fell dead. Carroll Williams was the brother of Jeff W. Williams of this county, who lives near Franklin. At the battle of Gettysburg July 2d, we had three color-bearers. The first one was Goodloe, who was wounded before he got to the house, and our recollections is lost his arm; the second was Carroll Williams, who was killed, and the third was A. J. Guice. Our brigade was composed of the 13th 17th 18th and 21st Mississippi Regiments, and the house and barn happened to be on the line of the 18th regiment when we made the charge.

* * *

WEEKLY CLARION-LEDGER
OCTOBER 22, 1903

PRIVATE FENTON L. FULGHAM
18TH MISSISSIPPI INFANTRY

[The writer] wound up his military career at Gettysburg in that memorable charge of Longstreet's corps, on the 2d of July, 1863. There is a small number of survivors of company E 18th Miss, who were participants and suffered the pangs of defeat and shard the glory of victory in that eventful period of our country's history. Among those were Capt. Billy Lewis, of Clinton, W. J. Green and Elisha Roberts of Copiah, John Brent of Raymond and Lewis Terrel "Old Red" who now lives between Jackson and Clinton. They are all about of Co. E. that I can recall, there may be others.

* * *

THE TOWN TALK
(ALEXANDRIA, LOUISIANA)
OCTOBER 23, 1914

SERGEANT WILLIAM M. FERGUSON
21ST MISSISSIPPI INFANTRY

Hero, Highly Praised By Gen. Lee, Is Dead

Winnsboro, La., Oct. 22. – The remains of Colonel Wm. M. Ferguson were laid to rest Wednesday.

Col. Ferguson was 71 years of age, and has lived in this parish ever since the war. He was a member of Barksdale's Mississippi Brigade. Longstreet's Corps, and in the last charge of Gettysburg saved the colors of the Twenty-first Mississippi from the very mouths of the enemy's guns, having received a personal letter from Gen. Lee commending his bravery and which Col. Ferguson cherished for life.

He was twice married, his second wife preceding him to the grave but a few weeks. He leaves several sons to mourn his loss and at the time of his death was worth considerable property in both town and country.

* * *

DAILY COMMERCIAL HERALD
VICKSBURG, MS
JULY 19, 1896

PRIVATE DENNIS COONEY
21ST MISSISSIPPI INFANTRY

Succor The Veteran

Somedays ago the C.H. published an appeal to the State authorities from Mr. Dennis Cooney, formerly a resident of this city and member of the 21st Mississippi Regiment, who is now disabled from the effects of a severe wound received at Gettysburg, and who cannot expect even the small relief to be afforded by the State pension, as he is not living in Mississippi. It is learned on most positive authority, that of members of the regiment, that he was in the service and was shot down, as he narrated, at Gettysburg, being the first man hit that day in in the regiment, if not in the brigade. No one of his old comrades who saw him fall supposed that he was alive and to hear from him after the lapse of thirty-two years was a great surprise to them. At the request and that of other sympathizing citizens the C.H. has opened a subscription list for his benefit and will receive a receipt for all money sent to this office for his benefit.

* * *

TIMES-PICAYUNE
NEW ORLEANS, LA
JUNE 3, 1866

The Remains of Gen. Barksdale. - Hon. E. Barksdale, formerly member of the Confederate Congress, from Mississippi, in a note to the editor of the *Clarion Standard*, refers to the recently published statements relative to the removal of the remains of his gallant brother, Gen. Wm. Barksdale, who fell at Gettysburg. He says "it is the intention of his nearest kindred to remove his body, as an early day, for consignment to its final resting place, at some appropriate spot within the limits of his own State, without taking the generosity of others."

* * *

VICKSBURG HERALD
MAY 20, 1867

ASSISTANT SURGEON A. T. HAMILTON
148TH PENNSYLVANIA INFANTRY
FROM LEWISTOWN PAPERS.

Camp On The Pike South Of Gettysburg, July 6, 1863.

I attended. Gen. Barksdale of Mississippi. He was shot through the left breast, and the left leg was broken by two missiles. He was brought by some staff officer to my temporary hospital. I have him what I could to relieve him.

He asked several times whether I considered his wound necessarily mortal, I told him I did. He desired peace, but only upon terms that would recognize the Confederacy. He was large, corpulent, refined in appearance, bold and his general, physical and mental appearance indicated firmness, endurance, vigor, quick perception, ability to succeed either as a politician, civilian, or warrior. He was a member of Congress under Pierce and Buchanan. He was asked about our strength, and was answered that heavy reinforcements were coming; said he "Militia under McClellan? He became delirious towards the last and said that Lee would show us a trick before morning; that we before knew it, "Ewell would be thundering in our rear."

He was dressed in grey jeans of their choice. His short roundabout was trimmed on the sleeves with gold braid. The cuff buttons with a star closed it. The collar had three stars of on each side next the chin; next his body was a fine cotton or linen shirt, which was closed by three studs bearing Masonic emblems. His pants had two stripes of gold braid, half an inch broad down each leg. Such was his attire, when the politician and warrior was laid on the soil of Pennsylvania to breath his last. - He was a Brigadier General in McLaws Division, Longstreet Corps, died July 3d at 2 o'clock A. M. and buried in a blanket at the house of Jacob Hammellbough, near Gettysburg.

This may be of use to some of his relatives who may perchance see it.

A. T. Hamilton
Ass't Surgeon, 148th P. V.

* * *

Excerpt from:
The Story of Our Regiment: A History of the 148th Pennsylvania Volunteers, 1904, p. 173

Alfred Thorley Hamilton
Asst. Surgeon, 148th PA at Gettysburg

I was the only medical officer at the left center front during the 2nd and 3rd of July until night of the latter day. I attended General Barksdale of Mississippi. He was shot through the left breast from behind, and the left leg was broken by two missiles. He was brought by some staff orderly to my temporary hospital. I have him what I had to relieve him. He asked several times whether I considered his wound necessarily mortal. I told him I did. He desired peace but only upon terms that would recognize the Confederacy. He was large, corpulent, refined in appearance, bald, and his general physical and mental make up indicated firmness, endurance, vigor, quick perception and ability to succeed whether as politician, civilian or warrior. He told me he was a member of Congress under Pierce and Buchanan. He asked about our strength and was answered that heavy reinforcements were coming. Said he, "Militiamen under McClellan?" He said that Lee would show us a trick before morning, that before we knew it Ewell would be thundering in our rear. He was dressed in the jeans of their choice. His short roundabout was trimmed on the sleeves with gold braid. The Mississippi button, with a star in the center, closed it. The collar had three stars on each side next the chin. Next his body was a fine linen or cotton shirt which was closed by three studs bearing Masonic emblems. His pants had two stripes of gold braid, half an inch broad, down each leg.

* * *

Excerpt from:
The Story of Our Regiment: A History of the 148th Pennsylvania Volunteers, 1904, p. 245

Joseph Wendel Muffly
Adjutant, 148th PA at Gettysburg

During the night of the 2nd as I approached the little house - now the field hospital—in front of the house lay General Barksdale, mortally wounded, his breast torn and one leg shattered by grape shot. Alternatively begging for water, which a drummer boy was giving him with a spoon, and cursing the Yankees, it was a most pathetic scene. "Bring me water, cold water," he would say. "When I am well I am a great lover of water, and now when I am shot all to pieces and burning with fever, I must have cold water." Then he would break out, "Yes, you think you have whipped us, but wait till morning and you will hear Ewell thundering in your rear."

* * *

EXCERPT FROM:
THE STORY OF OUR REGIMENT: A HISTORY OF THE 148TH PENNSYLVANIA VOLUNTEERS, 1904, PP. 542-43

HENRY MEYER
PRIVATE, CO. A, 148TH PA AT GETTYSBURG

Several hundred steps to the rear of our regiment was a small dwelling house with some outbuildings. In front of the house was buried General Barksdale of Mississippi. Comrade William C. Meyer (Private, Co. A) saw him soon after he was brought into our lines; he had cut off a fringe of gold lace from the general's coat collar, which he showed us. He also told us about the remark the general had made before he died, now famous in history, "Tell my wife I am shot, but we fought like hell." Sometime after the battle his remains were exhumed and taken to his people.

* * *

CONFEDERATE VETERAN
NOVEMBER 1896

J.J. HOOD
MERIDIAN MISSISSIPPI

BATTLE FLAG OF THIRTEENTH MISSISSIPPI REGIMENT

The Kemper Legion was the color company which bore this famous flag during this brilliant charge, and it was within a few feet of this historic emblem of the Confederacy, on the fire fringed heights of Gettysburg, where the gallant Barksdale, carrying his sword aloft at an angle of forty five degrees, led those steel nerved veterans on by boldness of action and heroic inspiration, till he fell mortally wounded near the enemy's guns.

Under its inspiring folds in this same dashing charge our esteemed fellow citizen, Mr. J.C. Lloyd, lost his arm, and there these venerable old veterans, Mr. F.C. Sinclair and Mr. Hunnicut, of our city, were badly wounded, and many other brave men baptized the soil of

Virginia [Ed. Note: Pennsylvania] with their heroic blood. This old tattered banner, flecked with blood, pierced by shot and shell, and hallowed by a thousand sacred memories and countless heroic deeds, strongly emphasizes the chivalric conduct of the fearless men who carved their fame under its waving folds.

This precious and priceless relic of the war tells of deeds that shall inspire our sons to a cultivation of the highest patriotism and that will be a stimulus to noble emotions. However cruel and relentless war may be, it has its virtues and compensations arousing a people to noblest deeds, to worthy sacrifices, and elevating them in every element of patriotic worth and manhood.

Strip them of these relics, histories and memories, and you soon degenerate their patriotic pride.

* * *

Excerpt from "Barksdale's Last Charge" by Fontaine Maury, read by Major J. J. Hood (possibly of the 21st. Mississippi) to the UCV Barksdale Camp of Meridian, Ms.

[Ed. Note: An article in *The Commercial Appeal* indicated that Fontaine Maury was a nom-de-plume for Brig. Gen. B.G. Humphreys. Maury was his wife's maiden name and Fontaine was the name of one of his children.]

THE DAILY NEWS (MERIDIAN MISSISSIPPI)
APRIL 9, 1901

"The first rays of the rising sun glinting from the mountain tops around about Gettysburg on the morning of July 2, 1863 discovered the enemy in full force drawn up in line of battle along the heights beyond the Emmitsburg turnpike. Far to the right rugged, impregnable Round Top rose in its awful majesty. To the left and farther off Culps Hill reared his grim visaged front, and the ridge between the two, as though fresh sown with the fabled dragons' teeth, literally bristled with bayonets.

"The enemy had been met in his own country, he had selected his own battleground, and with the precision of a soldier he had selected well Meade, however, had made one mistake. Instead of deploying his line along the top of the ridge between Round Top and Culps Hill, he made a detour into the valley about half way between the two near the Peach Orchard and Longstreet was ordered to attack this salient, and, swinging his right wing around Round Top, to fall upon the enemy's flank.

"Hood's division was ordered to make the assault on Round Top, and Barksdale to attack the salient and drive the enemy from the valley.

"About 3 o'clock the final order came 'The lines before you must be broken and the enemy must be driven from the field. Let every officer animate his men by his personal presence on the front line. Move forward at once and swing to the left.' The brigade was then in readiness to move.

"Humphreys commanded the right wing directly behind the Peach Orchard; Carter commanded the left wing' Holder and Griffin held the right and left center.

"Three hundred yards in front, passing hard by the Peach Orchard ran the turnpike from Emmitsburg to Gettysburg.

"Behind this, the enemy's first line was deployed, supported by the batteries on the elevation behind and in the Peach Orchard. The Washington Artillery opened fire on the line behind the turnpike. A puff of white smoke, a livid flash from the heights beyond, the thunder of a dozen batteries, echoed back in terrific splendor. The blow for liberty had been struck. Every face was blanched, every brow was knit. The carnival of death had been ushered in with the cruel roar of cannon, and the rattle of still more cruel musketry. The great decisive battle in the tremendous struggle for our independence was about to be fought. The fate of a struggling nation, whose prowess on the field of battle had dazzled a wondering world, whose army had never yet known defeat, hung on the issue of that day.

"Above the rattle of the muskets and the shrieking shells came the order to 'move forward.' 'Forward!' 'Forward!' rang down the lines from the regimentals, and the brigade moved out upon the mightiest battlefield this continent has ever known.

"In front of the brigade lay an open plane some three hundred yards in width across which they must charge the line behind the turnpike.

"Charge!" and then that yell which has baffled all description, born on the glorious field of Manassas, reverberated back from the valley of the Shenandoah, swelling in its tumultuous grandeur along that blood-bespattered plane which stretched alas, to Appomattox, it cannot be told, it cannot be written. Conception of hope, heroic inspiration of patriotism chivalric, the heart-swell of a storm-cradled nation that felt it burst forth like the glorious dawn of our morning and faded away like a song that is sung with the twilight of our hopes.

"On they charged. The heights beyond thundered with the savage roar of artillery. The turnpike blazed in one livid glare of death. The New York Zouaves, dressed in bright red uniforms, held the ground behind the turnpike, and were charged by Carter.

"From the Peach Orchard a battery of six guns belched screaming shells and shrapnel through the ranks of the advancing column, mowing them down like wheat before the sickle.

"The gallant Stamps, leading the left wing of the regiment, sword in hand, fell on the very threshold of the battle with a bullet through his heart. Like the wind breakers on a wide and wilder ocean they dashed across the open field against the turnpike. The Red Zouaves in front began to waver. 'Charge!' cries Carter, in his voice the ring of glorious victory. Another yell, another dash of the desperate whirlwind; and her struggle for supremacy, the red glare of another deadly volley, and the Red Zouaves gave way. Brave indeed was the stand they made, but they had fought 'not wisely but too well,,' for their shattered remnant reached the ravine at the base of the ridge the field was literally red with their slain. Carter, alas, lay dead on the turnpike.

"Humphrey's, in obedience to the order to swing to the left, swept the turnpike and pressed forward into the bloody Peach Orchard. Here he met the enemy's second line, and was sept by the deadly fire of the six-gun battery. The dashing Gibson, than whom no knightlier ever drew a sword, fell with a shattered thigh. 'Charge that battery!' and they charged it. - routed, riven."

Reeled the foeman's proud array;
They had struggled had and striven,
Blood in torrents they had given,
But their ranks dispersed and driven
Fed in suddenness away.

"Emerging from the Peach Orchard into the open plateau beyond he discovered that the brigades to Barksdale's right had been held in check by the enemy.

"Off to the right some 300 yards another battery was thrown into position to enfilade his column. In a flash "Right wheel, charge!" rang above the roar of the battle, and with the rattle of muskets, the clang of bayonets and the shouts of victory, Winslow's Massachusetts Battery fell a prize to the sons of Mississippi. Still the battle raged. The

Federal general, Graham, retreated across Plum Run and rallied his line behind a battery 1,000 feet away beyond the ravine.

"If that battery can be silenced and Graham routed again, the enemy's line will be completely severed. If severed, victory is ours. mayhap in this charge hands the history of the Confederacy. What a thought. What a moment. The command is given and the regiment moves forward to the ravine and under its shelter reforms in line of battle. Up the other side, and out upon plateau at the base of the ridge, they advance for the last and most desperate charge. Sickles rode to the front to encourage his men to resist the impending onslaught. Up the slope the Mississippians advanced, undaunted by the bursting shell or the screaming missiles.

"Thinned by the storm which swept down with such terrific fury from the ridge, the advance line staggered and began to waver. The awful crisis of that awful day had come. As on a thread hung the hopes of a struggling nation. Amid the shifting scenes of this doubtful battle the very course of history swung to and fro in the trembling balance. With the ebb and flow of this terrific battle rushed the tide of our country's hopes.

"Suddenly far to the left a wild shout of victory rings above the din of battle. Looking back toward the turnpike the majestic form of the heroic Barksdale, like the fabled god of battle, towering in the front of the advancing column was seen, leading the charge on the last line of the enemy at the ravine. The inspiration spread like contagion through the regiment, and with a yell that sounded to the very dome of heaven they rushed upon the enemy's line. As the autumn leaves are scattered by the blasts of winter, wo was the Federal line swept away by the charge of that hurricane. On, right on, until the fiery breath of the cannon was hot in their face, they pressed. The brave gunners who maintained that battery, with a fortitude worthy a better cause, never deserted their posts and died with their guns beside them. Sickles fell desperately wounded but escaped into his line. Graham, who had stood by his men till the last had fallen, fell into the hands of the Mississippians a prisoner of war. The last line was carried, the last battery was captured, the enemy's line was cut in twain.

"Thus it was the Twenty-first Mississippi regiment bore the stars and bars to the very furthest point reached in the enemy's line on that bloody field of Gettysburg.

"Holder and Griffin had swept across the turnpike in the pursuit of the Red Zouaves, and were now just making the last desperate assault upon the enemy's line. Barksdale stood on the turn-pike watching the movements of his heroes as they drove the enemy before them. His bright eyes flashed as though lit by a spark from the fires eternal. His face beamed with the glow of glorious victory. His thin lips curled with that brightly smile which meant defiance to the world. He stood the perfect picture of a perfect hero. Far to the right he had watched the gallant dash of the Twenty-first regiment and saw the enemy's line broken and scattered. He saw the Federal line reforming in his front across the ravine, and drawing his sword from its scabbard he sprang to the front, the flush of victory on his face, shouting, 'They are whipped. We will drive them beyond the Susquehanna.' Then it was the cheer went up which had lent the inspiration to Humphreys' wavering column. They charged the ravine as Mississippians were want to charge, but the Federal lines, reinforced, had rallied beneath its shelter and poured volley after volley, into his lines. His column, halted beneath the deadly fire, reeled to and fro like a forest beneath a tempest. The gallant Holder fell desperately wounded. Griffin fell. Fizer fell. Once more they dashed against the enemy. Waving his sword in the very thickest of the fray, in the last charge upon the enemy's last line.

Where swords were clanging, clashing. Where balls were crushing, crashing, the soldier, patriot, hero fell. Amid the wild roar of the cannon, the fierce scream of the shrapnel and clang and glitter of sabers; midst all.

The pomp and glorious circumstances of war, the great soul, the unconquered spirit of the immortal Barksdale winged its flight.

"No shaft marks the spot where he fell that day. The Federal authorities refused to all the point they reached to be designated by appropriate stones, but the story of that gallant charge is written upon the hearts of his countrymen and will be told in song and story as long as mankind read of the gallant deeds and heroism is a virtue.

"We are told that when the French made the attack upon the Austrian center at the battle of the Danube, Latour d'Auvergne, a private soldier, by his heroic example won the battle for his country. He fell in the charge, but ever after when his name was reached at roll call the sergeant would step to the front and answer, "Died on the field of honor." And so shall Mississippi in all the years to come, proud of the heritage he has left her, answer when the name of her knightliest sort is called, "Died on the field of honor."

* * *

EXCERPT FROM:
A MISSISSIPPI REBEL IN THE ARMY OF NORTHERN VIRGINIA: THE CIVIL WAR MEMOIRS OF PRIVATE DAVID HOLT

PRIVATE DAVID E. HOLT
16TH MISSISSIPPI INFANTRY

We wasted the second day of July in a fool proportion to repel a fancied attack when we were the attacking party. It angers me to this day, as I think of the cost of such actions which allowed a wonderful offensive opportunity to turn into a defensive one. We were furious then, our wrath aimed at Longstreet, in whom we had already lost confidence.

Night came at last and put a stop to the intolerable heat. All night we could hear the artillery and the infantry coming up and digging in, over on Cemetery Ridge, and yet, not even our skirmishers fought. There had been fighting going on both to our right and left.

During the night, Brother Joe came from the right where his field hospital was located in a cut in the road. He easily found me, and we lay in the trench and talked for some time. He brought something to eat and recent news from home. To him, I was little Stevie, I have never grown up, so far as he is continued even to this day. He was deeply moved by the risk to which I was exposed, and he looked forward with dread for me on the morrow. There would be fearful casualties, he said.[19]

* * *

"PEEL, WILLIAM H. COMPANY C, DIARY 1863-1865"
MISSISSIPPI DEPARTMENT OF ARCHIVES AND HISTORY
JACKSON, MISSISSIPPI

LIEUTENANT WILLIAM H. PEEL
11TH MISSISSIPPI INFANTRY

Gettysburg "2d"

We were aroused at an early hour by our Major, who came to warn us to rejoin the reg't immediately. We were agreeably surprised to find we had been spared the anticipated ducking. Calling in the outposts,

19. Thomas D. Cockrell, and Michael B. Ballard, eds., *A Mississippi Rebel in the Army of Northern Virginia: The Civil War Memoirs Of Private David Holt*, (Baton Rouge, LA: Louisiana State Press, 1995), 195-96.

we returned to bivouac, where we found the wagon train already in motion, + the reg't formed. We took our position in line, + the reg't moved off with the train. After an advance of a couple of miles we met Gen'l Davis coming after us. We gave him three hearty cheers in acknowledgement of our approval of the conduct of himself + the balance of the brigade, on the day before, + were conducted by him in the direction of the present bivouac of the rest of the brig.

We passed the brigade Hospital on the way, where there were friends of many members of our reg't suffering from their wounds of the day before. Soon after passing the Hospital, we left the wagons + moving off to the left some mile perhaps, came to where the balance of the brigade were. The smallness of the reg'ts compared to their size on leaving us the morning before, together with the number of mangled forms we had just seen at the Hospital told in language quite unmistakable, the cost at which had been bought, any glory that might be expected to redound from their achievements of the day before.

Ranks being broken, friends met with cordiality that might be expected under the circumstances + we were entertained with many stories of gallantry in the late engagement. Heard that the color sergeant of the 2d reg't had, in a hand to hand engagement, had his flag taken from him, while another member of the same reg't had, in a smaller contest, carried off the colors of a Yankee reg't. + learned with gratification that the 55th N. C. reg't –one of our brig, had deported itself in this its first battle, in a very commendable manner.

Meantime the battle raged furiously as the distance of a couple or three miles from us. The fitful outbreaks of musketry that ever + anon burst upon us from the right showed the stubborn determination with which the "Old Bull Dog," as Longstreet has vulgarly, but with some propriety, been called, was hurling his brave men upon the naturally all but inaccessible heights it had been apportioned him to carry if possible. Ewell, on the left was far from idle, while the continuous roar that came from the centre told plainly that Hill's work was not child's play.

Several batteries of our corps were within distinct view, at the distance of about a mile. They played with great rapidity on the enemy's lines. I watched the firing for some time, +, to get the better view, climbed to the top of a small tree. Long red streams of flaming breath flashed forth, then quickly came the thunder peal, as if, like the discordant yells of the Hindoo at the Cossack, to drown the dying groans + shrieks of its unfortunate victim.

During the evening I walked over a portion of the field of the day before, witnessing all the horrifying spectacles that visit the "field" the day after a battle. Implements of war were scattered in every direction, while here + there lay horses + men in every conceivable degree of mutilation. Our wounded had all been taken to the Hospitals, + those who were still there had been collected into a grove of forest trees. There are perhaps a few stages of suffering of which the imagination may conceive, that were not here represented.

I saw one poor fellow, lying stretched on his back. He must have been lying down when he was struck, for a minnie ball had struck him near the top of his head + appeared to go directly in. There was quite a quantity - a handful it seemed to me - of his brain that had issued from the wound + was lying on the ground. The man appeared perfectly unconscious, but was breathing freely, then at least 24 hours from the time he was shot.

When I returned to the bivouac, I found that two or three of my Co. had been to Gettysburg - a couple of miles off - a portion of which our troops now held; having driven the enemy partially out. They brought back some molasses + several other articles, stating that the town had been deserted by the Citizens, the doors of houses being left open, +, all sorts of goods scattered over the streets. It will be wondered perhaps why they were so moderate in their demands. The reason is simple enough: two + a half years experience had taught them that a soldier in active campaign, having a blanket, a change of underclothes + the necessary rations, was supplied with every thing with which he had any business.

Late in the evening the brigade was formed + we moved up a couple of miles to the right, + taking our position in line, were ordered to rest, probably for the night.[20]

* * *

YAZOO HERALD
MAY 15, 1908

FIRST SERGEANT WARREN D. REID
11TH MISSISSIPPI INFANTRY

I am indebted to two veterans "who wore the gray" for valuable papers on the great battle of Gettysburg - one of them Mr. Warren D. Reid of Houston, Miss. He belonged to Company H, 11th Mississippi Regiment. The other one is Mr. McNamara of Wilkerson county, whose home is now at Beauvoir. He belonged to the 21st Mississippi Regiment. These two soldiers could not have written of a more

20. Also found in: Ellen Sheffield Wilds, Ed., *Far From Home: The Diary of Lt. William H. Peel 1863-1865*, (Bloomington, IN, 2009,) 53-75.

desperate battle, or any other where Mississippi troops behaved with more superb courage than here at Gettysburg, called by some writer, "The Waterloo of the Confederacy." Gen. John B. Gordon said Gettysburg and Appomattox fix the boundaries of the Confederacy's decline and death. At Gettysburg its sun reached its zenith and passed its meridian; at Appomattox it went down forever. The mighty hosts had met in the fierce death grapple. The supreme struggle at Gettysburg was at hand. Thee it was the 21st Mississippi Regiment bore the stars and bars to the farthest point reached on that bloody battlefield. We learn from one who knew, that "the Seventeenth United States Regulars lost 62 per cent in killed and wounded in the Peach Orchard, and upon the ridge beyond. The state of Massachusetts has erected a magnificent monument to mark the spot where Bigelow's Battery was annihilated. There is no stone there to tell this story, but the fact remains that the regiment that swept through that Peach Orchard on Gettysburg's crimson field and up the height beyond, mowing down the 17th U. S. Infantry and destroying that battery, was led by a Mississippian, a son of Claiborne county."

* * *

EXCERPT FROM:
A CONFEDERATE SOLDIER'S HISTORY OF THE ARMY OF NORTHERN VIRGINIA IN 1863, (VICKSBURG, 1896)

SGT. CHARLES E. HUTCHINSON, 48TH MISSISSIPPI, COMPANY H

The Diary of Charles E. Hutchinson

July 2 – We have moved to the front to support battery. Sent out skirmishers and threw up temporary rifle pits for our protection. In the evening our battalion was sent forward to support skirmishers, who were driving the foe. Advanced to within three hundred yards of battery, and could have taken it, if we had been supported, I think. Stayed out until dark and came back to pits. Our adjutant was killed. Loss to our company – (wounded) – Kain, shot in hand; McRaven, shot in ankle; Gibson, in the head, and Wiley, slightly in the hand. It seems impossible for men to advance in such a storm of shot and no more be killed. The shelling was terrific.

* * *

THE VICKSBURG EVENING POST
JULY 19, 1913

PRIVATE FRANK H. FOOTE,
CO. F "CLAIBORNE VOLUNTEERS," 48TH MISSISSIPPI

Writes home after the Battle of Gettysburg

Hagerstown, Md., July 9, 1863

My Dear Sisters and Brothers:

Through Divine influence I have again been preserved through another hard fought battle, and most thankful do I feel to Him for having been so good to me. I suppose ere this you have heard of the battle and its result, therefore I will not enter into any details, except that which concerns my own company and regiment, and with the other Claiborne companies. The battle was the most obstinately contended one of the war, and for the first time in 10 or 12 battles has the old battalion been forced to about face, and leave the ground. Such fighting has never been witnessed before on this continent, and I pray to God that it may never be so again. The Yankees had one of the most formidable positions in the country I reckon, and all our efforts to dislodge them proved futile. [Ed. Note: July 2, 1863] Our regiment and the 48th Georgia charged and captured a 11 gun battery, but was forced to leave it, as we did not have any support near us. In the charge our company suffered more than any other one in the regiment, viz killed: George Stampley (cut into [sic] by a shell; Clay Johnson (leg shot off with a shell). Wounded, Warren Richmond, slightly; Tom Applegate, L.S. Thompson (in arm and hip); Sol Labenburg, arm slightly, and W.J. Seaman. Seaman also missing, not heard a word of him since making the charge. I hope he is not hurt. I think he is a prisoner. The regiment sustained a loss of 42 killed, wounded and missing in about 1-2 hour time we were engaged. The brigade did not suffer very much. John Shields and Bob McAveen were slightly wounded (Co. G 16 Miss.) None I believe in the Guards. The battle lasted 3 days, and on the 3rd night of the battle we (the army) silently withdrew and fell back to this place, to get ammunition, as we never had enough to protect ourselves in case of an attack. You must not judge by this that we are defeated. No, far from it, as all accounts, both our own and the Yankee's say it was a drawn battle. The Yankee papers admit a loss of between 35,000 and 40,000, while ours will amount upwards of 20,000. Lt. Walker had two narrow escapes during the battle, a 24 lb. solid shot rolled down on his shoulder, not hurting him any though, also a minie ball went through his blanket he had on his shoulder, making 18 holes through it. Several other of the boys were slightly wounded by spent bullets. We lived fine while in Penn. and drew rations of chickens, vegetables, butter, apple butter and soup, cherries there was no end to them, such cherries, the size and color of muscadines, some like a large red plum, others of various sizes and colors, oh such delicious fruit. I wish you were here to enjoy them with us. Everything we got we paid for in Confederate notes, none refusing to take pay.

We are now lying in camp doing nothing but eating, cooking, and sleeping. We are looking for a battle almost at any hour, and do not care when it comes, as gen. Lee says that he is going to conquer peace in the Yankee's own land, and let them feel some of the horrors of a civil war. In conclusion I bid an affectionate adieu, and remain as ever your affectionate brother.

Frank

My love to one and all of you, and I will write more fully next opportunity.

My regards to all inquiring friends. All send theirs to the family.

* * *

AUGUSTUS L.P. VAIRIN, DIARY, JULY 1, 1863, MDAH, JACKSON
1ST SERGEANT AUGUSTUS L.P. VAIRIN, CO. B, 2ND MISSISSIPPI

We moved about one mile to the right of yesterday's battle and rested all day ? gathering arms etc. from our bivouac. We could see the Battlefield of this [day]. I was struck in the head with a glancing ball which addled me for the rest of the day.

* * *

EXCERPT FROM:
"ROBERT AUGUSTUS MOORE DIARY (VOLUME 1)."
UNIVERSITY OF MISSISSIPPI CIVIL WAR COLLECTION
ROBERT A. MOORE, PRIVATE CO. G, 17TH MISSISSIPPI

Thursday, July 2, 1863, in line of battle on the battlefield

Had a desperate encounter with the enemy this evening for 2 hours. Drove them before us for 1-1/4 miles but were forced to fall back for lack of support. Captured several batteries & stand of colors. Our loss was heavy, in the Regt. 223 killed & wounded, in our Co., 29. Several of them were my dear friends. Every man acted the hero. Miss. has lost many of her best & bravest sons. How thankful should all be to God who have escaped. OH! The horrors of war.[21]

21. Also found in: James W. Silver, ed., *A Life For The Confederacy: As Recorded in the Pocket Diaries of Pvt. Robert A. Moore* (Jackson, Tenn., 1959) p. 153.

Private Robert A. Moore of the 17th Mississippi Infantry

(Esther Cooper Cannon, Louisiana Historical Quarterly)

Corporal L. Purnell of Company I of the 11th Mississippi Infantry
Purnell was wounded on July 2, 1863 at Gettysburg

(Liljenquist Collection, Library of Congress)

Chapter Four

Gettysburg Day 3 — Friday July 3, 1863

Following unsuccessful assaults on both Union flanks on July 2nd, late in the morning of the next day Lee determined to launch a frontal assault on the Union center as soon as his Confederate forces could be marshaled. The attack would be spearheaded by the fresh Virginia division of Longstreet's First Corps under flamboyant Major General George E. Pickett — three brigades commanded by Generals Lewis Armistead, Richard Garnett, and James Kemper — supported by forces from A.P. Hill's Third Corps — Heth's Division (now commanded by Brigadier General James J. Pettigrew) and two brigades from Major General Dorsey Pender's division under command of Major General Isaac Trimble. Popularly known as Pickett's Charge, this assault is more accurately named the Pickett-Pettigrew-Trimble Assault. As finally deployed the attacking column from south to north consisted of Kemper's and Garnett's brigades supported by Armistead; then Archer's brigade commanded by Colonel B.D. Fry, Pettigrew's brigade under Colonel J.K. Marshall, Davis' Mississippi Brigade, and Colonel J.M. Brockenbrough's Virginia Brigade under Colonel R.M. Mayo. Alfred M. Scales' brigade under Colonel W. Lee Lowrance and J. H. Lane's brigade were deployed behind Archer's and Pettigrew's brigades in support, leaving Davis' and Brockenbrough's brigades *without any support.* On the Confederate extreme right and rear were Wilcox's Alabama Brigade and Perry's small Florida Brigade.[1]

As part of Heth's division, Joseph P. Davis' Mississippi Brigade saw desperate action on the extreme left flank of the charging column. Barksdale's Mississippi Brigade (now under command of Colonel Benjamin Humphreys), reduced to half-strength with command structure in tatters, was in no condition to continue the fight. Surprisingly, Posey's Mississippi Brigade who had been scarcely engaged on July 2nd, and only saw skirmish action again at the Bliss farm during the morning of July 3rd by the 12th Mississippi, were ordered back to support Pegram's artillery on Seminary Ridge where they remained throughout the third day.[2]

---

1. Coddington, *The Gettysburg Campaign*, 489-90.
2. *OR* 27/2, 634.

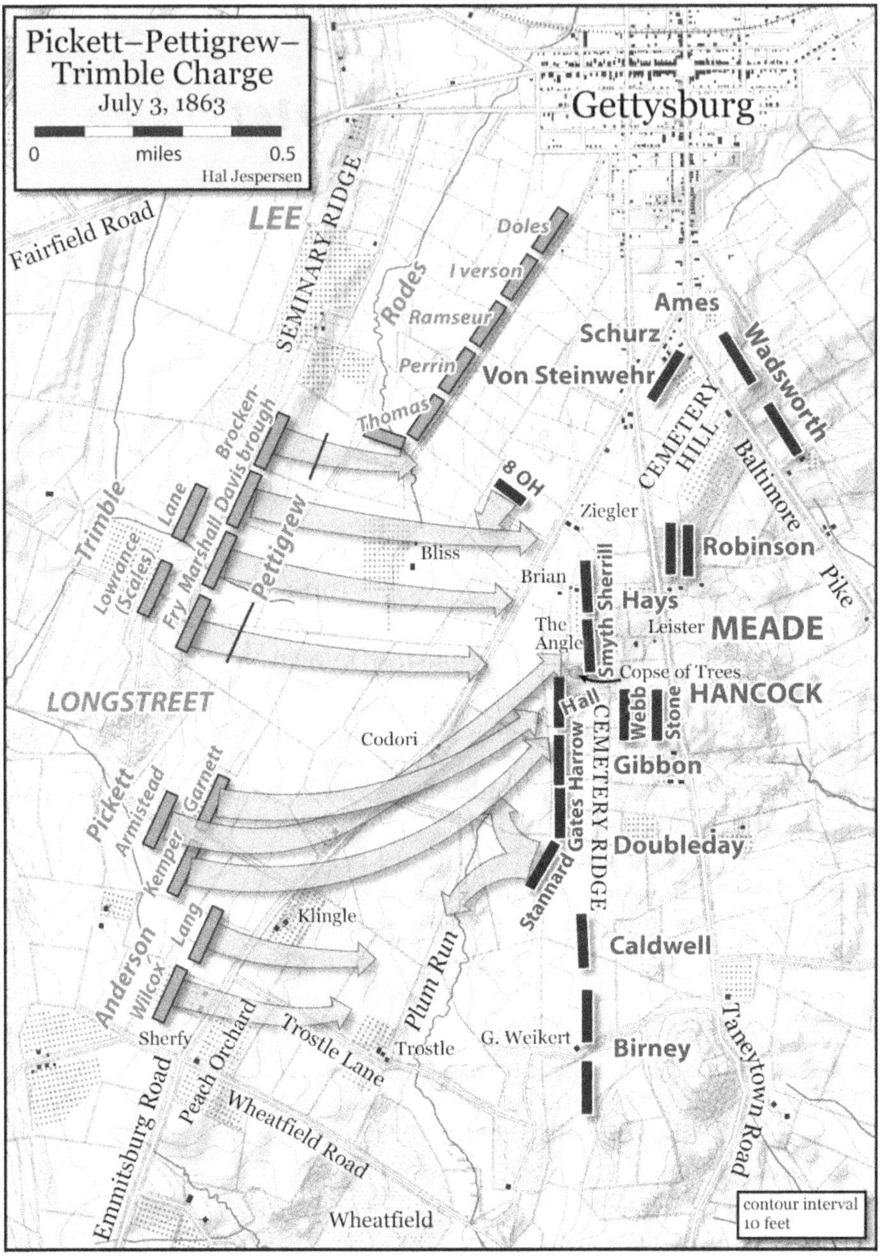

The mere presence of Brockenbrough's brigade, and especially Davis' Mississippi Brigade, in the attacking column has long been questioned by historians and authors. In fact, one member of the 2nd Mississippi later wrote: "When General Lee saw the cut-up and worn-out condition of our men on the third day in the field, he remarked that they should not be there." Regarding Brockenbrough, Edwin

Coddington observed that, by July 3rd, severe campaigning and losses of 300 on July 1st had reduced this once "splendid outfit" to the size of a regiment. Weakest of the four brigades under General Pettigrew, surprisingly it held the crucial position covering the left flank of the attacking column. Regarding Davis' Mississippi Brigade, Author T.P. Williams has commented:

> **In view of their losses on the 1st it is astonishing to think they should have participated at all, yet their deployment near the critical and extremely vulnerable left flank strains all tactical sense. Was no one in the high command aware of their shattered condition? Were they considered expendable? Perhaps the answer lies within the tragically confused command structure of July 3rd. Except for Pickett, nearly every key figure associated with the attack was substituting for someone else.**

Coddington went on to speculate whether deployment of these two badly mauled brigades was "by accident or intent.:"

If Lee wanted the two Pender brigades [Scales and Lane] to advance en echelon on Pettigrew's left instead of following behind his right, either he did not make himself clear to Longstreet, or Lane misunderstood Longstreet's instructions. On the other hand Lee and Longstreet might have decided to place Pender's brigades behind Pettigrew's right so as to increase the weight of the attack near the center and ensure the penetration of the Union line near the Clump of Trees. In such an arrangement Pettigrew's two left brigades [Davis and Brockenbrough] would become expendable, for their role would be to absorb enfilading fire and flanking attacks from Cemetery Hill.

Coddington's words proved prophetic for Davis' Mississippians on July 3rd![3]

At 1 p.m. two signal guns of the Washington Artillery of New Orleans at the Peach Orchard fired in quick succession commencing the Confederate bombardment. More than one hundred and twenty cannon arrayed along the two-mile Confederate line on Seminary Ridge erupted with a deafening roar. Soon over eighty Federal cannon along Cemetery Ridge replied in like manner resulting in the greatest cannonade on the North American continent. Posted behind Pegram's artillery on the crest of Seminary Ridge Davis' Mississippians, like Barksdale's the day before, endured collateral damage from the Union fire. The 11th Mississippi suffered two killed and twenty-one wounded.

3. *Samuel W. Hankins, Corporal 2nd MS, Simple Story of a Soldier: Life and Service in the Second Mississippi Infantry, 1905*, 50-51; Coddington, *The Gettysburg Campaign*, 490-91; T.P. Williams, *The Mississippi Brigade of Brig. Gen. Joseph P. Davis*, 105.

Sergeant Jeremiah Gage of the 11th Mississippi Infantry

(H. Grady Howell, Jr.)

Second Sergeant Jeremiah Gage, a popular young member of Company A, the University Greys, was mortally wounded by a shell fragment and was taken to the nearby field hospital. Dr. Legrand J. Wilson recorded: "A piece of shell had struck the left side near the stomach, tearing away the tissues, a rib, and the spleen and fractured the left forearm, lacerating it terribly." Second Lieutenant Daniel A. Featherstone, also from Company A, was killed outright as described by Second Lieutenant Andrew J. Baker:

> **While we were lying down on Seminary Ridge at Gettysburg, just behind our batteries, during the most unprecedented cannonade ever witnessed on earth, and**

> **just before we went into the charge on Cemetery Heights, Featherstone was struck by a cannonball and thrown several feet in the air, his head being split wide open.**

This bombardment continued for about one hour, shrouding the scene with thick smoke and acrid sulphurous fumes which, coupled with eighty-seven-degree temperatures and no air circulation, caused further suffering to Davis' Mississippians lying prone in the woods.[4]

About 2 p.m. the cannonade ceased and an eerie hush fell over the smoky field as these Confederates tensed in anticipation. Lieutenant William Peel of Company C spoke for the entire regiment with this entry in his diary:

> **A silence as awful as the thunder of the minute before had been ... settled itself around us. All ears were strung for the command we knew full well must follow soon. 'Attention' rang loud & clear, along the line. In an instant all hands were upon their feet, and the line was dressed. The ashen hue than lingered upon every cheek, showed the accuracy with which the magnitude of the task before us was estimated, while the firm grasp that fixed itself upon every musket, & the look of steady determination that lurked in every eye, bespoke an unflinching resolution to 'do or die'. Alas! how truly was that resolve sustained.**

Davis formed his battleline with the 55th North Carolina on the right and the 11th Mississippi on the left with the 42nd Mississippi and 2nd Mississippi in the center. Confederate artillerymen cheered as the infantry emerged from the woods with banners flying. General Davis later reported: "The order to move forward was given and promptly obeyed." Due to the alignment of the opposing battlelines the Confederate left, including Davis' brigade, was the first to come into view of the Federal defenders. Momentarily awed by the grand spectacle, and husbanding ammunition, Union gunners did not open fire. Davis later confirmed: "Not a gun was fired at us until we reached a strong post and rail fence about three quarters of a mile from the enemy's position, when we were met by a heavy fire of grape, canister, and shell, which told sadly upon our ranks." Tumbling over this "slab fence" these determined Mississippians pressed on. Confusion reigned supreme within Brockenbrough's Virginia Brigade. Accounts vary but

4. LeGrand J. Wilson, *The Confederate Soldier*, 122; Edward J. Stackpole, *They Met At Gettysburg*, (Harrisburg, PA, 1956), 254; Stubbs, *Duty*Honor*Valor*, 420-23; Andrew Jackson Baker, "Veterans Camps and Inquiries," *Confederate Veteran*, Vol. VI, 1898.

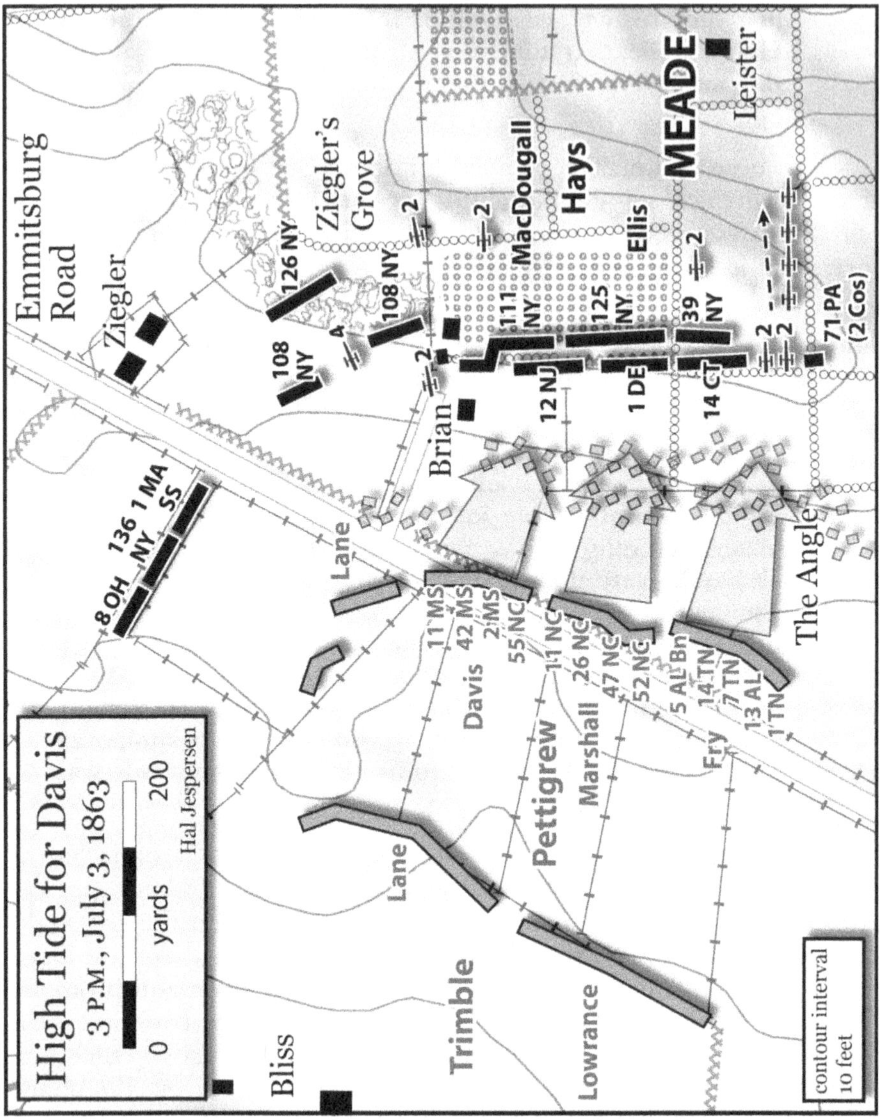

this contingent failed to advance initially with Davis' Mississippians, struggled to catch up, and passed beyond the Bliss farm. Now under severe fire from their front and left by Major Thomas W. Osborn's thirty-two cannon and deadly musketry from the 8th Ohio under Lieutenant Colonel Franklin Sawyer on their left flank, Brockenbrough's brigade stopped short of the Emmitsburg Road and melted away to the rear, thereby exposing the 11th Mississippi to this same deadly conflagration.[5]

5. "Lt. William H. Peel Diary," July 3, 1863, MDAH; *OR* 27/2, 650-51; Stubbs, *Duty*Honor*Valor*, 424-27.

Private Thomas F. McKie of the 11th Mississippi Infantry

(Maud Brown)

The Union troops directly in front of the approaching Mississippians and behind a low stone wall were from Brigadier General Alexander Hays Third Division of Hancock's II Corps including the 12th New Jersey of Colonel Thomas Smyth's 2nd Brigade, and the 39th New York, 111th New York, and 125th New York of Willard's 3rd Brigade now under command of Colonel Eliakim Sherrill. The 108th New York and the 75-member picket

Captain John H. Moore of the 11th Mississippi Infantry

(Confederate Veteran Magazine)

reserve of the 126th New York advanced to the right and formed on Davis' left flank. As Davis' brigade reached the Emmitsburg Road these Union regiments unleashed several devastating volleys into their front and left flank and these Mississippians went down in heaps—but steadily pressed on! Seventeen-year-old Private Tommie McKie a member of the University Greys and already a veteran soldier was mortally wounded in the abdomen and died the next day. Young Color Sergeant "Billy" O'Brien carrying the regimental battle flag of the 11th Mississippi was killed instantly by a Federal shell. His likeness is immortalized on the life-sized monument to the 11th Mississippi on the west side of Confederate Avenue, dedicated in 2000, and located where the regiment formed for the July 3rd assault. Their sister regiments to the right were also taking horrendous casualties. Lieutenant Colonel David Humphries of the 2nd Mississippi was killed while Colonel Hugh Miller of the 42nd Mississippi was mortally wounded. Indeed, as Author T.P. Williams has noted: "By now every field officer in the brigade [except for General Davis] had been shot down. The remaining company officers were dropping fast." Mississippi veteran William A. Love recorded the desperate scene:

> **No one seemed to be in command. It had become a soldier's battle, in which the Southerner's watchword, "The grave of a hero or victory," was being gloriously exemplified. Captain John Moore, of Company A, "University Greys" (Eleventh Mississippi) was in front, facing the regiment and trying to close up the fearful gaps being cut in the line, when Lieut. A.J. Baker, of the same Company, shouted, "For God's sake, John, give the command to charge!" —"No," replied Moore, "I cannot take the responsibility," whereupon Baker himself gave the command and the thin line rushed forward to the stone wall as individuals rather than as an organization.**

Baker was subsequently wounded and captured within ten feet of the stone wall.[6]

Three members of the 55th North Carolina–Lieutenant T.D. Falls, Co. C; Sergeant Augustus Whitley, Co. E; Captain Fletcher Satterfield, Co. H— reached the stone wall which was recessed 47 yards farther east than that pierced briefly by General Armistead and his Virginians. Slatterfield was killed while Falls and Whitley were forced to retreat. Regimental Adjutant Thomas Holliday and Acting Adjutant for Davis' brigade, Captain William Magruder, (brother of Major General John B. Magruder) were both killed nearby. To the north at the Brian barn, framed on both sides by the stone wall, the small remnant (estimated at about 40 men) of the 11th Mississippi reached its furthest advance–formally recognized by the Gettysburg National

6. Gottfried, *Brigades of Gettysburg*, 167, 173-75, 180; Laino, *Gettysburg Campaign Atlas*, 368-72; Maud Morrow Brown, *The University Greys: Company A Eleventh Mississippi Regiment Army of Northern Virginia 1861-1865* (Richmond, 1940), 46; Stubbs, *Duty*Honor*Valor*, 428; T.P. Williams, *The Mississippi Brigade of Brig. Gen. Joseph R. Davis*,102; Love, "Mississippi At Gettysburg," 44, 46.

Captain Fletcher Satterfield of the 55th North Carolina Infantry

(North Carolina's Regiments by Clark 1901)

Park Commission as the Confederacy's true "High Water Mark." Lieutenant Peel, charging forward with his sword in his right hand and pistol in his left, recorded the final moments here:

> **The flag staff was now cut in two midway the flag ... now dangling, in graceless confusion, from one corner, still pushed fearlessly, upon the stone fence. Thirteen of our Reg't had concentrated upon the colors, as if to constitute ourselves its guard. We were some yards in advance of the line, & now found ourselves within about thirty yards of the stone fence. Immediately before us was a small framed house—about twenty feet square—the farther end of which joined the fence. Springing forward, we secured its shelter, gaining at the same time, a position, within twenty five feet of the Yankees behind the fence. The boys betook themselves to the work before them in good earnest. A number of shots were fired, which must have proven very fatal, as the distance was so small.**

The sacred battle flag of the 11th Mississippi was recovered by a first sergeant of the 39th New York, the Garibaldi Guards. Unable to locate a senior officer, Captain John Prince of Company D, The Neshoba Rifles observed strong reinforcements advancing from the Union rear, and shouted the order to retreat. Lieutenant Peel and several others were soon captured at the Brian barn.[7]

Just like General George Pickett, Brig. Gen. Joseph R. Davis escaped unscathed and his precise whereabouts during the assault remain unknown. Similar to Pickett, Davis was overcome with emotion at the horrendous Confederate losses he witnessed that afternoon. Private David Holt of the 16th Mississippi, Posey's Mississippi Brigade, encountered General Davis during the confused retreat of his shattered brigade. Holt later recorded the encounter:

> **Soon the frazzled remnant of [Davis'] brigade came staggering back through the smoke singly or in small groups. The organization of the brigade had been completely broken up. General Davis, with his sword in his hand, came slowly down our line. I knew him personally and went up to him and asked, "General Davis, where is your brigade?" He pointed his sword up to the skies, but did not say a word, and stood there for a moment knocking pebbles out of the path with the point**

7. T.P. Williams, *The Mississippi Brigade of Brig. Gen. Joseph R. Davis*, 104; Love, "Mississippi At Gettysburg," 47; Stubbs, *Duty*Honor*Valor*, 429; Peel, Diary, MDAH, July 3, 1863.

Monument to the 11th Mississippi Infantry flagbearers on the Gettysburg National Battlefield.

The monument depicts William O'Brien of Company C who was the first of eight flagbearers of the 11th killed or wounded during the famous charge on July 3, 1863.

(Tim Fulmer, GettysburgSculptures.com)

Mississippi monument on Gettysburg battlefield

(Joseph L. Owen)

of his sword. He could not talk, and neither could I. He walked on and I went back to my place.[8]

The 11th Mississippi suffered greater casualties than any of the fifteen Virginia regiments under Pickett, and its 87% casualty figure (i.e. 341 killed/wounded/captured out of 394 present) surpassed that of all Confederate regiments engaged that day. Topping the ten companies, the University Greys endured an astounding 100% casualty rate for the 31 members present. In recognition of the sacrifice, fortitude, and accomplishment in reaching the Confederate "High Water Mark," the final marker placed within the GNMP in 2000 is the memorial honoring the 11th Mississippi Infantry Regiment located just in front of the stone wall at the Brian farm.

Private Tom Wilkins of Company E, the Prairie Guards (one of four brothers in the assault) was concussed by a shell and suffered a fractured right leg. As Tom struggled back to McMillan Woods he encountered what he described as the "saddest sight I have ever seen." In recording his heart-rending description Wilkins provided a suitable epitaph for the University Greys:

8. Pvt. David Holt, Civil War Memoir, 198.

> All of the little black servants owned by the University Greys were standing in a row wringing their hats in their hands, with tears streaming down their dusty cheeks. Each boy cried out his master's name, but no master answer. **The University Greys had ceased to exist. They were all gone — gone to glory up on Cemetery Ridge**. (emphasis added)[9]

Thus ended the three-day bloodbath at Gettysburg—with Mississippians contributing mightily to the Confederate attacks on each day. Mississippi veteran William A. Love succinctly summarized Mississippi's contribution with these words:

> Davis's brigade ... fought its way unsupported into the First Union Corps on the first day, losing heavily in killed, wounded and captured. Barksdale's victorious advance on the second day proved abortive for lack of timely support. Davis, again on the third day, was ingloriously forsaken in the assault and left to advance under both front and flank fire, and then cursed for having 'given way.'

Only Posey's Mississippi Brigade was spared a prominent role due to a breakdown in communications (despite their ardor for combat and through no fault of their own), merely acting as skirmishers and protecting artillery batteries on July 2nd and 3rd. The final word goes to an unnamed Mississippi Veteran of the battle who wrote to the *Weekly Democrat Times* of Greenville Mississippi in 1887 in response to a previous article written by a fellow Mississippian who neglected to mention Mississippi's role at Gettysburg:

> If there is anything in the martial history of Mississippi of which she can boast; if there is one bright page in all her annals; if there is inspiration for the youth of our state to be gathered from the heroic deeds of her sons at all, certainly Gettysburg furnishes the theme.
>
> **The history of Gettysburg with no mention of Mississippi, is the story of the Alamo without Crockett, Thermopylae without Leonidas.** (emphasis added)
>
> Had Mr. Toombs consulted the records at Washington City, where he has been for the past few years, **the most casual perusal of the list of the killed and wounded would have shown him the name of Mississippi written all over that fateful field in the blood of her bravest sons.** (emphasis added)[10]

9. Stubbs, *Duty*Honor*Valor*, 420, 425, 431, 435, 437. Contains a listing by name/rank/company for all casualties from the 11th MS. See pp. 903-05.
10. Love, "Mississippi At Gettysburg," 51; "A Mississippi Vet," *Weekly Democrat-Times* (Greenville, MS), June 21, 1902.

* * *

CONFEDERATE VETERAN
AUGUST 1893

W. GART JOHNSON
18TH MISSISSIPPI COMPANY C

REMINISCENCES OF LEE AND OF GETTYSBURG

[W. GART JOHNSON - Orlando, Fla., July 18, 1893]

It was on the morning of the 3rd of July, 1863 at Gettysburg. On the evening before Hood and McLaw's divisions of Longstreet's corps, on the right wing, had driven the enemy from all his positions on the open plain to the stronghold of Cemetery Ridge. My company (C, 18th Mississippi), with others, was occupying the extreme front picket line in direct range of the sharpshooters. We were in the edge of an apple orchard. Adjutant Harmon, of the 13th Mississippi, and I were hugging a pile of rubbish, anything to hide behind, that we had thrown together, when Gens. Lee and Longstreet??-----on foot, no aides, orderlies or couriers, fifteen or twenty steps apart, field glasses in hand-----came walking past us, stopping now and then to take observations. They were arranging, as we soon found out, for the famous charge of Pickett's division. As Gen. Lee halted in a few feet of us, knowing the imminent danger he was in, one of us said, "Gen. Lee, you are running a very great risk." At that moment the searching minnie was cutting close to him, showing that he was the mark aimed at. He went on with his observations as calm and serene as if he was viewing a landscape. A few minutes afterward we heard him say to Longstreet, in substance, "Mass your artillery behind that hill," pointing to a ridge just in our rear "and at the signal bring your guns to the top of the ridge and turn them loose." It put us to thinking of what would become of us-----the picket line. We could not leave our posts; we were in plain view of the enemy, without protection except from small arms; we had no utensils with which to throw up earthworks. We knew the shells from our guns would go over us, but those of the enemy! Well, spades or no spades, we went into that ground quicker than you would think. We were like the fellow after the ground hog, it had to be done. Bayonets, pieces of board, any thing to get out of sight. Two or three to a hole, and we went in like gophers. That was the grandest and at the same time the most terrible artillery duel I ever witnessed. Think of it. There were sixty-five (I was told) of our own pieces on that one spot, and more on another portion of our line, all firing as fast as they could, and the cannon of the enemy replying. I don't know how long it lasted. When it stopped on our side Pickett's division charged! They had to march over us. Doing nothing myself, I had time to look. It was one of the grandest sights ever mortal eyes looked upon. It makes me shudder now, as I see the shells plow through the ranks of that gallant band.

* * *

Copy in 17th Mississippi Folder, GNMP

Private John S. Henley, 17th MS, Company A

3d Day's Battle At Gettysburg

When morning came, having lost nearly half our men the day before, our company was sadly depleted in numbers. As soon as it was light we had to relieve our pickets who had been on duty all night. We moved cautiously so as to find the enemies' line. We slowly moved until about 150 yards from the enemy. Just then, three Yankees arose in front of me, and raised their guns, but I was ready, and got the first shot. Two bullets passed over my head. We stayed out there all day, and kept up a continual fire.

Pickett's Division came up during the night. General Lee commenced making preparations to attack Culp's [Cemetery] Hill. About 12 o'clock we opened fire on them with about 100 guns, and the artillery fire was the heaviest I heard during the entire war period. I was out on the picket line and a shell passed over me every second. About one o'clock, the Infantry moved forward in what is known as "Pickett's famous charge."

* * *

Vicksburg Evening Post (Vicksburg, MS) 1896

Sgt. Charles E. Hutchinson, 48th Mississippi, Co. H
Soldier Diary

July 3 – We were awakened this morning by the picket firing, and soon after the cannon opened up, and then we had a lively time of it. About twelve lines of battle were formed, and advanced about 2 o'clock, but the enemy was too strongly posted. Before our lines advanced, our cannon opened, and were replied to by the enemy; and such a rattling of shell I never heard before. It was one continued roar for about two hours. I thought I had been under shell at Manassas, Sharpsburg, and Fredericksburg, but I never saw shelling until today. A great many horses of our battalion were killed. Do not know our loss in front. Expect it was heavy.

* * *

WILSON, THE CONFEDERATE SOLDIER, 120-21

LEGRAND J. WILSON
SURGEON, 2ND MISSISSIPPI INFANTRY

July 3rd, 1863

Early in the morning I was ordered to my command. I had been constantly engaged the previous day and half the night attending to the wounded of the first day's battle, then I slept three hours, hastily ate my breakfast and hurried to the battle field. I found Holt and Shields waiting for me and, together, started out to find a suitable place for a field hospital. After walking ourselves down, we selected a large barn as the best place we could find for our wounded, with water at hand and plenty of hay and straw for bedding. This was a typical Pennsylvania barn ? lower story built of stone. So we had three walls of stone between ourselves and the battlefield, which would protect us against bullets, grape and canister, and here we prepared for the coming storm. And caring for our wounded, which consisted chiefly in extracting bullets, legating bleeding vessels, checking hemorrhages in different ways, splinting fractured limbs so that the poor sufferers could be sent in ambulances, to the real hospital for appropriate treatment. This hospital was usually two or three miles in the rear of the battlefield. To apply the term hospital to this field station was really a misnomer.... Here often had the sad pleasure of writing down the last messages of love and affection to the dear ones at home, whispering words of consolation and hope into the ears of the dying.

Toward noon the artillery commenced firing, and for three hours we had pandemonium. Think of three-hundred pieces of artillery firing as fast as the men could load them coupled with the bursting of shells, and possibly you may form an idea of the noise. The air seemed to be in a state of vibration, and produced a very strange feeling in one's head.

The cannonade over, the order, "Forward," was given. The boys in gray advanced over the hill, and the contest was begun.

* * *

LETTER, H. Q. BRIDGES TO MRS. V. G. ARMISTEAD, JULY 26, 1913. GAVE FAMILY COLLECTION, SPECIAL COLLECTIONS, UNIVERSITY OF MISSISSIPPI ARCHIVES.

SERGEANT JEREMIAH 'JERE' SANDERS GAGE
11TH MISSISSIPPI INFANTRY

[Ed. Note: One of the earliest casualties on July 3rd; mortally wounded during the cannonade]

Mrs. V. S. Armistead
2615 Royal Street, N. O.

My Dear Madame: -

On my return from attendance at the recent Gettysburg Celebration held July 1, 2 & 3, inst., I found a letter awaiting us from my brother at Nolansville, Miss., with enclosure of a contribution to the *N. O. Times Democrat* of Sunday, June 23, signed Flo Field under the caption: -

"Jere Gage, Hero of Gettysburg," with reproduced photograph.

My brother suggested that no doubt you and your sisters, whose names appeared in the article, doubtless would appreciate a letter from me, telling the facts and circumstances of how Jere died, as I was there at the time. He was one of the noblest boys I ever knew, and was my college mate at the state university, my mess mate during the war as members of the University Greys, later known in military annals as Company A, 11th Miss. Vols. Inf., Davis Brigade. Heth's Division, A. P. Hill's Corps, in the Army of Northern Virginia. I was at the university from '58 to '61, he having graduated in '60. So, it will be seen I was an intimate personal friend and a longtime associate of his. We were together during the war from the time of enlistment in May '61 up to Gettysburg, where on July 3, '63 he yielded up his splendid young life on that famous battlefield. I'm sending you herein a copy of the inscription on the "Marker" erected by the government at the exact spot where our regiment was engaged at the time he received his mortal wounds, which I myself witnessed, as I was within a few feet of him when he was struck. We were in the battle the first day until dark; the second day we were not in action; but in the forenoon of the third day we were in a strip of woods by the Emmitsburg road on Seminary Ridge about one mile south of the town. We were at the time under the most galling fire that ever came out of a cannons mouth, the very ground underneath us seeming to shake from the plunge of solid shot and the detonations of bursting, shrieking shells, which appeared to rake and search out every square yard of the ground we occupied. My remembrance at this long distance of time is that Jere was in a sitting posture while the most of us, except, of course, the officers, were lying down. Presently a 24 pounder shell exploded immediately in front of us, and Jere called out that he was struck, one of the shells hitting almost tearing his left arm from the socket, and another fragment hitting him in the abdomen, either of which terrible wounds would have been fatal. Jim Dailey, also one of our mess, being litter bearer that day, at once gathered him up, and placed him on the litter and born him away to the field hospital about 200 yards away to the south of us, where it was that Dr. Holt saw him and attended him. Of course none of us could leave ranks but what I know as to how he died I was told by Jim Daily that evening about night, when I was taken to the field hospital to have my wounds attended to, for having been in the charge on the enemy's stone wall breastworks in the afternoon. I was myself seriously wounded upon the hand by a Minnie bullet. Jim told me that Jere died soon after being carried to the field hospital and said that Jere asked him to "Bury me in my old shawl, Jim," and which he did. It was the fashion of the boys at the university to wear shawls in lieu of top coats, and it was this shawl that Jere had carried all through the war in place of a blanket that he requested Jim to make his winding sheet. It was about noon of the fatal day when he was wounded, hence it was that Dr. Holt said he was the first to be brought to the hospital. It was about 2:30 P.M. when we were ordered to advance against the stone wall across the valley; it was there I was disabled, and was not taken to the hospital until about dark. I have not seen Jim since, and have heard that he was dead. In my recent visit to Gettysburg I met only three of the old regiment, but none of my old company "A," as with a possible few exceptions they are all out of action now, having gone to "Land of the leel;" but together with those three we went on the 3rd day to the point indicated by the "Marker" as being the place where we stood in the forenoon of that day 50 years before, and all agreed it was the exact spot, and it was a hollowed spot for us, for we were fairly overwhelmed with a flood of memories of the long ago.

On a separate sheet I'm enclosing you the inscription on the Marker, which you will find interesting because it tells with perfect accuracy what part we took as a regiment in that great battle. Similar ones at various other places also record the action taken by other troops on both sides. Alas, dear Jeremiah Saunders Gage, Peace to thy Soul.

"Green grows the grass above the.
"Friend of my early days.
"None knew the but to love thee.
"none named thee but to praise.

Very Respectfully,
H. D. Bridges

Copy of Inscription on Marker at Gettysburg Battlefield.

C. S. A.
Army of Northern Virginia
Davis' Brigade, Heth's Division A. P. Hill's Corps.
55 N. C.; 34 Miss; 11th Miss; 42nd Miss. Regts. Inf.
July 1. Formed line west Herr's Tavern and crossing the Run
At 10 A. M. Joined later by 11th Miss previously on duty guarding train.
The brigade fought until the day's contest ended.
July 2. Lay all day west of the Run and at evening took position near here.
July 3. In Longstreet's assault this brigade formed the Left Centre
Of Pettigrew's Division, and advanced to the stone wall south of the Bryan Barn,
Where, with regiments shrunken to companies and field officers all disabled, further
Effort was useless.

July 4. After night withdraw and began the march to Hagerstown.

Present First Day about 2,000
Killed 100; Wounded 717; Missing 500, Total 1397.
Company A

There were present the first day	31
Killed during the 3 days	14
Wounded	17

This addition is made by me.

* * *

LEXINGTON ADVISER
JULY 11, 1913.

SERGEANT JEREMIAH "JERE" SANDERS GAGE
11TH MISSISSIPPI INFANTRY

Death of Jere Gage at Gettysburg

The following article from the *Commercial Appeal* will be read with much interest by citizens of Holmes County as Jere Sanders Gage was a resident of the county and has several relatives still stationed here:

Jere Sanders Gage, of Richland, Holmes County, Mississippi, my classmate, class of 1860, University of Mississippi joined the University Greys, Eleventh Mississippi Infantry at Oxford, and was killed in Pickett's charge at Gettysburg, July 3, 1863.

I am sure that the memoranda herewith enclosed, will be of intense interest to all readers, and respectfully request a place for the copies in the *Commercial Appeal*. Very Truly,

James Goodloe

* * *

COPY OF PART OF DR. JOSEPH HOLT'S ARTICLE OF JUNE 22, 1913,
IN THE NEW ORLEANS TIMES-DEMOCRAT,
DESCRIBING THE DRAMATIC AND SIDNEY LIKE DEATH OF JERE GAGE.

"Dear Madame - In my reminiscences of Gettysburg in last Sunday's Times - Democrat, the princely fellow and favored son of the Eleventh Mississippi, the first to arrive on a litter was Jere Gage. His body is buried where he dated his letter, 'on the battlefield," and I trust, will ever remain, for the field of Gettysburg, is his monument. That you, his sister, has announced his name from this time hence the story of his departure will live in cherished memories of the Southern people, and as part of the common heritage of "magnanimity and valor of American manhood in even balance.' He needs no tablature in marble' brass or bronze. That you are his sister I profoundly congratulate you, madame, and myself in being the medium of information. Very sincerely,

(Signed.) Dr. Joseph Holt."

"How One Hero Died."

"The first to arrive on a litter was a princely fellow and favored son of the Eleventh Mississippi. ...Keeping everybody close to the ground, I turned to him and he pointed to his left arm. I quickly exposed it, and found that a cannon ball had nearly torn it away between the elbow and the shoulder. I made some encouraging remark, when he smiled and said: 'Why, doctor, that is nothing; here is where I am really hurt,' and he laid back the blanket and exposed the lower abdomen

Dr. Joseph J. Holt, Surgeon of the 2nd Mississippi Infantry

(New Orleans Times Democrat)

torn through from left to right by a cannon short, largely carrying away the bladder, much intestine and a third of the right half of the pelvis; but in both wounds so grinding and twisting the tissues that there was no hemorrhage, I then surveyed him personally observing the tender devotion of his bearers, and I saw a singularly attractive creature. Through his deadly pallor I could detect a sunburned blond who in health would show a strong and ruddy countenance; a large head, with a tousled shock of reddish golden locks, like a mane, with the musculature and form of an athlete. Deferentially polite, there was something singularly, self-confident and manly about him, answering distinctly the descriptive marks of that Shepherd, the younger Jesse, at about the time he chose him five smooth stones out of the brook, for he was ruddy and withal, a beautiful countenance, and goodly to look to.'

"Without the slightest change of voice he asked: Doctor, how long a time have I to live?" A very few hours, I replied. Doctor, I am in great agony, let me die easily, dear doctor. I would do the same for you.' His soul peered from the depths of his blue eyes in appeal of anguish that cut me to the heart, and I replied: 'You dear noble fellow, I will see that you will die easy."

"No word or detail of this scene has faded from my memory. There was no thought of the dramatic; it was dreadfully genuine, and naturally spontaneous in the unconscious creating and acting of a grander tragedy than he might ever hope to play.... I called for, and my hospital knapsack bearer, Jim Howell, quickly handed me a two ounce bottle of black drop, a concentrated solution of opium, much stronger than laudanum. I poured a tablespoon of it into a cup with a little water, and offered it: but, a thought flashed into my mind, and, withdrawing the cup, I asked: 'Have you no message to leave?' It startled him, and in a low, moaning wail he cried: 'My mother! Oh, my darling mother! How could I have forgotten you? Quick! I want to write. By that time all were crouching under the low shelter of the embankment, including several of his regiment, were crowded around, oblivious of their own injuries and weeping silently. His last message, I took my seat on the ground, close beside him to steady his head, his elbow in the hollow of my right hand to support and steady his arm, and a pencil slipped to his hand. Jim Howell had provided the sheet of paper held in the smooth lid of the hospital knapsack improvised as a desk. He wrote rapidly – all of this transpired in haste – murmuring to himself the words audible to me, for I looked the other way. He began with place and date – "On the Battlefield, July 3, 1863. 'He wrote little more than half a page, into which he poured with vehemence his soul of tenderest love, never faltering for a word: and a message toward the last, with a name that he wrote silently, conscious of the presence of strangers; but the message was too personal and sacred to him for me to trespass; for it was holy ground. The last line he softly repeated aloud. "I dip this letter in dying blood. With that he turned down the blanket and seizing the letter, pressed a back of it upon his oozing bloody wound, and handed it to me, giving his mother's address and begging me to be sure she got, letter .... I saw that she got the letter. I arose from the ground and had him supported, when he turned to me with a reminder of my promise, and his hopeless pain. I handed him the cup and he waving it, saying: 'Come around, boys, and let us have a toast. I do not invite you to drink with me, but I drink the toast to you and to the Southern Confederacy and to victory.

He quaffed to the last drop, and returning the cup to me, said: 'I thank you.' I rushed off to attend to other wounded. In another hour, passing hastily, I lifted the cover from his face, to find him sleeping painlessly.... I passed again and laid the cover from his face, to find him sleeping painlessly .... I passed again and laid the cover from his face, to find his face, to find the reincarnated soul of Sir Galahad had taken its flight into triumphal ascension to Him who instituted and consecrated the Holy Grail. Oh, the excruciated pathos, and very agony of the glory! His death unsurpassed in tenderness of love," in philosophical resignation, in courage and will in sacrifice to self, it is unrepassable, even that of Socrates."

* * *

LETTER, JEREMIAH GAGE TO PATIENCE GAGE, JULY 3, 1863
GAGE FAMILY COLLECTION, SPECIAL COLLECTIONS, UNIVERSITY OF MISSISSIPPI LIBRARIES.

SERGEANT JEREMIAH S. GAGE
11TH MISSISSIPPI INFANTRY

[Ed. Note: Final letter from 'Jere' Gage stained with his blood]

Gettysburg Penn

July 3rd

My dear mother,

This is the last you may ever hear from me. I have time to tell you that I died like a man. Bear my loss as best you can. Remember that I am true to my country and my greatest regret at dying is that she is not free and that you and my sisters are robbed of my worth whatever that may be. I hope this will reach you and you must not regret that my body can not be obtained. It is a mere matter of form anyhow. "This for my sisters too as I can not write more. Send my dying release to Miss Mary." you know who. J. S. Gage Co. A, 11th Miss. (This letter is stained with my blood.) (1)

* * *

FAR FROM HOME: THE DIARY OF LT. WILLIAM H. PEEL 1863-1865, ED. ELLEN SHEFFIELD WILDS, (BLOOMINGTON, IN, 2009) 53-75; WILLIAM H. PEEL, COMPANY C, DIARY, 1863-1865, MDAH, JACKSON, MISSISSIPPI.

"PEEL, WILLIAM H. COMPANY C, DIARY 1863-1865"
MISSISSIPPI DEPARTMENT OF ARCHIVES AND HISTORY
JACKSON, MISSISSIPPI

LIEUTENANT WILLIAM H. PEEL
11TH MISSISSIPPI INFANTRY

Gettysburg '3d'

The night of the 2d, broken in upon occasionally by a report from the artillery of the wakeful enemy, + at one time by a sudden, tremendous rattle of musketry, the result of a charge by a portion of Ewell's command, showing the continued wariness of the Federals to be by no means superfluous, wore slowly away, + the morning of the 3d at length dawned upon us. We remained in our position until 9 or 10 o'clock in the fore noon, when the artillery immediately in front of us, open fire, in reply to which the enemy threw over a number of shells, several of which, striking in the ranks of our reg't suggested the propriety of a removal of a few hundred yards to the right. The firing ceased after a few rounds, + all was comparatively quiet, for a few hours.

About 11 o'clock we moved some five or six hundred yards to the left, passing Posey's brigade on the way, when I met several friends in the 19th, for a moment, + took position in a skirt of grim woods We were halted + commanded to rest in place.

Seeing by surrounding circumstances what we were soon to look for, we distributed ourselves to the best advantage, behind tree stumps + c. some digging holes with spades that were scattered around + which Gen'l Pender troops had used during the morning for the same purpose, making ourselves safe as possible, with the means + time at our command, + prepared to await in the coming of the anticipated storm.

That moment was one of awful sublimity. For two days the conflict had raged with dreadful slaughter + with unabated fury. The balance seemed so far to be in our favor. We had driven them on nearly every part of the field. The ground that our portion of the line now occupied was about three miles in advance of where we had struck them on the 1st. But every hour brought the enemy fresh troops + additional supplies of munitions, while all of Gen'l Lee's forces had reached the field by the morning of the 2d; his entire command numbering from the best I can learn, a little less than 100,000 men. All had been engaged, with a very small exception, during the two days previous, to a greater or less extent, + all were more or less exhausted in consequence. Recent rains too, having swollen the Potomac so as to render it impracticable, had cut off Gen'l Lee' supplies of ammunition, thus pending all the greater importance upon the occurrences of the evening now fast approaching. The invasion was a grand design, + worth the adventure. Defeat + a precipitate retreat across the Potomac, on the one hand, + the routing of the Yankee army: the overrunning of a vast extent of country: the acquisition of immense supplies of all kinds. The capture of numerous towns + cities, among them the Capital of the U. S.; +, in short, the termination of the war, on the other, seemed poised in the balance to be tilted by the result of endeavors of the next few hours. But had Deity now hung out that golden Scale the Scale in which all created matter once was weighed; + in which the import of passing events has since been tried – the same by which was shown the weakness of the Price of Spirits damned, in the threatened contest with Gabriel + his angel band, when the latter had apprehended him in the perpetration of his devilish designs upon our sleeping parents – made bare to the contemplation of the human eye; there pending from the yard-arm of the ship of eternal ages, with the great events of this eventful day arrayed against each other, the result, even then had probably not been changed: so nearly horizontal must the beam have stood.

A few yards to our front; arranged in one continued zigzag line, generally on higher ground than that occupied by us, One Hundred + Fifty brazen + iron monsters looked with frowning hate across the field to a distance of one mile upon a like number of deadly enemies; similarly arranged, but on more elevated positions, all charged to their muzzles, with their most deadly missels for the occasion.

"Twas noon; + the moment for action had arrived. The signal guns sounded along the line, +, all on a sudden that long-to-be-remembered duello began. Our guns lead off. Almost at the same instant, along the whole line, there burst a long loud peal of artificial thunder forth, that made the ground to tremble beneath its force. The enemy's guns, in the same breath, as it were, took up the contest, + sent a tremendous shower of shot + shell whizzing over us. The storm now became deafening. Peal on peal, in one continuous roar, belched forth the dread artillery, while the very atmosphere around us, became vocal with the humming of the death-dealing demons that whistled by.

For three long hours this infernal storm prevailed. For three long hours the musicians of earth, with instruments well-tuned, played aloud this hell-born air, while all the spirits of the infernal regions danced + laughed with happy glee, at this triumph of their infamous cause, over the Holy banner of King Emmanuel.

In the hottest of the cannonading I heard a shell strike in the right of the reg't, +, turning over, as I lay upon my back, I looked just in time to witness the most appalling scene that perhaps ever greeted the human eye. Lt. Daniel Featherston, of Co. "F.", from Noxubee county, was the unfortunate victim. He was a large man – would have weighed perhaps Two Hundred pounds. He was lying on his face when the shell struck the ground near his head, + in the ricochet, entered his breast, exploding about the same time + knocking him, at least ten feet high, + not less than twenty feet from where he was lying.

About 3 O'clock the artillery firing ceased. A momentary of the men resuming their places in line, + silence – a silence almost as awful as the thunder of the minute before had been = settled itself around us. All ears were strung for the command we knew full-well must follow soon.

"Attention" rang loud + clear, along the line. In an instant all hands were upon their feet, + the line was dressed.

The ashen hue that lingered upon every cheek, showed the accuracy with which the magnitude of the task before us was estimated, while the firm grasp that fixed itself upon every musket + the look of steady determination that lurked in every eye, bespoke of an unflinching resolution to "do or die." Alas! how truly was that resolve sustained. "Forward" now resounded along the line, from right to left.

Moving forward as few yards, we passed our line of artillery, + a little further on, we crossed a low stone fence + found ourselves in a large open field, extending far to the right, left + front, rising with a gradual slope in the latter direction.

Far over the field to the front, at the distance of perhaps a little less than a mile, + near the top of the ascent there lay a long dark line, parallel with our advancing front, which a stranger would probably not have understood, but which we recognized as a stone fence – something very common in this section of country. Nor were we blind to another important fact: that behind this fence was sheltered a line of Yankee infantry. Just in rear of this fence, on the more elevated knolls, was ranged the enemy's scowling artillery.

The peaceful contemplation of the scene, however, was destined to be short lived. Our debut in the field was greeted with a broad-side from the long line of artillery that sent a storm of screaming, howling shells across the field, that burst + tore the timber behind us in a frightful manner. Volley after volley resounded, as broadsides came forth, until they subsided into one unbroken roar, louder than the thunder-peals of Heaven. Shells, screaming + bursting around us, scattered their fragments + projectiles in every direction. "Steady boys" – "+ slow" – "Don't break yourselves down by running."

Alas! that such commands should ever have found their way into the mouths of Confederate officers.

We were now advancing in the face of a perfect tempest of maddened shells that ploughed our line + made sad havoc in our ranks. As we moved onward we were greeted, as we came successively within range of these less far reaching, but far more destructive projectiles, with showers of grape + canister, + at the distance of about two hundred yards the infantry opened on us from behind the stone fence. Pressing onward, we returned the fire. Our line was now melting away with an alarming rapidity. It was already reduced to a mere skeleton of the line of one hour ago. Still on it pushed, with a determination that must ever be a credit to the Confederate soldier.

Four brave men had already fallen under the colors of our Reg't, + now the fifth bore them aloft, + rushed boldly forward to embrace, if need be, the fate of the other four. The flag staff was now cunt in-two midway the flash but without one moment's pause, the never-flinching little Irishman (Geo Kidd), his flag now dangling in graceless confusion, from one corner, still pushed fearlessly upon the stone fence. Thirteen of our Reg't had concentrated upon the colors, as if to constitute ourselves its guard.

We were some yards in advance of the line, + now formed ourselves within about thirty yards of the stone fence.

Immediately before us was a small framed house – about twenty feet square – the farther end of which joined the fence. Springing forward, we secured its shelter, gaining at the same time, a position within seventy five feet of the Yankees behind the fence. The boys betook themselves to the work before them in good earnest. A number of shots were fired, which must have proven very fatal, as the distance was so small.

Thinking the line rather a long time coming up, I looked to the rear.

The state of my feelings may be imagined, but not described, upon seeing the line broken, + flying in full disorder, at the distance of about one hundred + fifty yards from us.

What was to be done. A momentary consultation decided Lt. R. A. McDowell + I were the only officers with the party. I being the senior, the responsibility, if needed there were any, devolved upon me. There were but two alternatives: to surrender, or become the "flying target" of a thousand muskets. We preferred the former + in a moment more a white flag floated from behind the corner, around which the moment before our accurately aimed muskets had belched their deadly contents into the ranks of the enemy. An old serg't came out + took charge of us, + ordered us through the gate that was open on the left of the house through the gate that was open on the left of the house. As passed through, all unarmed of course, a Yank soldier brought down this musket, + with its muzzle right at the breast of one of our party was on the point of firing. I cringed for the safety of my brave comrade; + shuddered at the thought of seeing him thus butchered, but just at this critical juncture, our serg't sprang forward, knocked up the musket +, with a word of reproach asked the soldier if he did not see that the men had surrendered.

On passing the line, we were surrounded by a crowd of soldiers, all of whom were anxious to take charge of us. (It is a mighty good thing to get to take prisoners to the rear, especially when the front is as well heated up as that at Gettysburg was.)

We were hurried under a strong guard. Our retreating line had, in the meantime, repassed our line of artillery, which immediately resumed its fire. Shells came screaming through the air, + began to burst around us. They flew thicker, + faster as the greater number of pieces opened, + I saw several very narrow escapes.

Notwithstanding there was a line of skirmishers a few yards in the rear of the line of battle, to prevent men from leaving the ranks, there were a good many scattered over the field, behind rocks, + hillocks, in ditches + such other places of security as they could find. I saw three or four fellows peeping from behind a huge rock, to see what was going on, when, all unawares, a large shell struck the rock within a few feet of their heads. I suppose they all thought themselves killed, for they tumbled to the ground + laid quite close for some time. The shell rebounded + flying off a hundred yards to the left, narrowly missed ploughing its way through a squad of prisoners + their guard. For half a mile we were thus exposed to the fire of our own guns. Hobbling along as best we were able - most of us wounded - hurried at every step by our impatient guard, we at length reached the Baltimore Pike, where we found the second line of battle.

Geo. Kidd +I, being each of us wounded in the leg, refused longer to keep up the pace at which we were going where upon a single Irishman was left with us, + the balance proceeded on their way: I could not help thinking it somewhat strange that two honest sons of the Emerald Isle should be thus engaged in deadly combat against each other, both in foreign hands, + neither possessed of much stimulant in the shape of worldly affects. Kidd's remark to the guard showed his mind to be running in the same channel. "I am surprised," said he, "to see so many Irish men in your army."

"And I've been thinking the same of ye," said the guard.

"I think my position altogether natural,' replied Kidd. "But for you who have seen your mother land ground into the dust by the nason [?] heel of the usurper, to engage in a war for the subjugation of the South, seems to me to be inconsistent." "Ireland! Ah, it is for her dear sake that I am fighting now, answered the other.

"Truly then," replied Kidd, "misery loves company. Because you have seen your native land usurping under the British yoke, you propose to help the United States reduce the Confederacy to the same deplorable subserviency merely that there may be another nation in the world as miserable as your own." "Not that, not that," returned the guard, "but you know in Ireland we looked hopefully to the day win the U.S. would interest herself in freeing us from that bondage. And, as a preservation of the strength of the Federal Government, I to see the Union stand." Kidd answered, "The killing of a million of its men is a bad way to preserve a nations strength."

We here met a staff officer, who, noticing my portfolio – one so contrived as to roll up into a small round package – asked me if it were glass: I told him it was not; + informed him what it was, whereupon he asked to see it, saying it would be a good thing to carry papers in.

I handed it to him, stating that was what it was intended for, but I hoped the Officers of the Confederate army were not such fools as to entrust papers, of any importance to a man who would have no better sense then to carry them in so predictable a place as that.

He examined, + returned it to me, when we proceeded on our way.

We at length reached a point of concentration for our part of the line. Here we found about a thousand of our country-men, who had been captured during the day or the two days previous. We were turned into the pens, with the balance of the herd, + now found ourselves under guard in an open field.

Meantime there were several surgeons going around to see after the wounded, + each in his turn, was as well cared for, as could, under the circumstances have been expected. My turn came about dusk. I was glad of its arrival, for I had begun to experience some pain from my wound.

Clouds hung threateningly over us; +, without the prospect of shelter, we dispose of ourselves as comfortably as circumstances would permit, + waited for the coming of the showers.

There were small squads + large squads being constantly brought in, + by night-fall our number was increased to sixteen or eighteen hundred.

Upon examination, the Doctor pronounced the wound of no material consequence, being only a flesh cut, + not very deep. Binding it up, + within the bandage, he cautioned me against unnecessary walking, as such exercise would be very likely to inflame a wound in the knee + passed to the next man.

Genl Patrick, the Provost Marshall General, came upon the ground about this time.

"What's the news Gen'l, shouted some of our party, guessing at his rank, by his trappings + the staff accompanying.

"Bad news – bad news for you + myself too. Gen'l Longstreet, who was a particular friend of mine, + a nice man, was mortally wounded, fell into our hands + has died, said Genl' Patrick. "Sorry to hear it" was the distrustful reply. The Gen'l now waved his hand, for silence, + having partially restored order, addressed us thus: Prisoners, you are here now in my charge; quite a large number of you: I guarantee you the kindest treatment the nature of the case will permit, so long as you conduct yourselves in a becoming manner. If, however, there should be any attempt upon your part to escape me, woe be unto you. My splendid cavalry is at hand, armed + ready for action, + in number almost equal to your own, + in case of any disturbance among you, they shall be ordered to charge you, cutting + slashing right + left, indiscriminately."

A battalion of the "splendid cavalry" maneuvered around us a little during the evening, accompanied by a lady in a handsome riding habit.

Darkness had now drawn his somber veil effectively over this awful field – miles in extent – from which the life-blood of thousands of dead + dying proclaim aloud the depth of the mighty abyss of depravity into which man has fallen. Oh, that the magic spear had been but a little sooner applied to the creeping Tempter, that his fiendish designs might have been thwarted, himself banished from the realm of bliss, + our first parents left to the everlasting enjoyment of their Heaven-on-Earth.

One by one had my companions-in-misfortune, overcome by the Herculean exertions of three days of deadly combat, sunk upon the wet ground, into the sweet sleep of soldiers, to snatch a moment of respite from all his anticipated horrors. All, immediately, around me was comparatively silent, as I sat upon a rail; pondering my situation, the prospects before me, + aspect of affairs generally. Along the Baltimore pike - about a hundred yards from us – the trundling of artillery, the rumbling of wagons + the heavy tromp of troops, moving in the direction of Baltimore, were the only sounds that broke through the silence of the night. The column passed on, with little noise, having much of the appearance of a retreat, until the year had passed, + silence again prevailed broken, now, only by the tramping of a sentinel on his beat, but a few yards from me.

Thus left to themselves, my thoughts wondered for a while to various sections of the country + to different incidents + occurrences of life, but tiring of the flight, soon settled themselves down for repose among the "oved ones at home," + when I thought of the fact that it was for their sakes that I had taken upon myself the burden which but now seemed to bear heavily upon me, I thanked God that I had been so constituted as to be able to endure, than I have yet been called upon to undergo.

My revelry was brought to a sudden termination by the challenge of the sentinel, halting the relief guard, + I was brought to a true sense of my condition, to find myself "still sitting on the rail."

The picture which now presented itself was by no means a flattering one. No prospects of supper, save a piece of raw bacon + a cracker. Fortunately, however, I felt no hunger. No shelter other than the broad canopy of Heaven – not even the sheltering foliage of a forest oak to shield me from the falling mist; not even one single fire, to dispel the blackness of night that gathered so thick around, nor bed nor covering save the wet ground below + the lowering clouds above. Feeling the web of Orpheus gently twining itself about me, however, I laid me down, on the bare wet ground, + with a half filled haversack as a pillow, soon forget in sleep, the new-born troubles that seemed to spring up in my path before me. (1)

* * *

A Mississippi Rebel in the Army of Northern Virginia: The Civil War Memoirs Of Private David Holt, Thomas D. Cockrell,, and Michael B Ballard, Eds. (Baton Rouge, LA, 1995) 196-200.
Excerpt from: David Eldred Holt Papers
Private David E. Holt
16th Mississippi Infantry

At last day broke and we looked across the valley and saw the strong fortified position of the enemy bristling with artillery and blue with Yanks. The fighting began both on our right and left. As Lee's line was about fifteen miles long, we had no means of ascertaining the results.

Some time after noon, the artillery on Seminary Ridge opened fire. The Yanks replied in kind. The whole battler field soon became enveloped in smoke. Through the smoke came a shower of round shot and shells. The roar got deafening. Our artillerymen stood by their guns gallantly, while the commanding officer of the batteries rode back and forth just to the rear of the guns. He had his hat in his hand, and while we could not hear what he said, we knew that he was encouraging his men. He would dash off to the right and be lost to our sight in the smoke, then come back and ride up to the guns in front, waving his hat and shouting and then dash off to the left, and again we would lose sight of him. It became to us a matter of intensely interesting speculation as to how long he could keep that up.

But we had matters of our own to attend to. Some of the men put rails in the breastworks in front of them. We warned them not to do that, and called their attention to the disaster wrought in the Yankee ranks at Chancellorsville by this same method. But it proved slow work to loosen hard ground with a bayonet and throw it up with a frying pan. He had only two spades in our company and only one pick, and while you got a spade when your time came, the time seemed long in coming. [The boys tried] to hasten things by using rails.

Then solid shot hit the breastwork near the bottom of the bank turning the whole pile of dirt over them. They were buried out of sight while the sold shot was lying on top of all; rails, dirt and men. We rushed to the rescue of the buried men and had to work in feverish hast in order to get the men out before they suffocated. Besides, the cannon booming in front of us and the shells bursting around us had a most accelerating effect, an effective but unpleasant nerve tonic. Two of the men were near the suffocation point and all were full of dust within as well as without. Water was scarce, and it was with difficulty that we drained enough out of the combined canteens of the company to give them some moisture. Then we had to work like Trojans to mend the gap in the breastwork. All this time the sun took advantage of our plight and fried us to a "doneness" like "spareribs" [and we had little in our ribs to spare].

The wounded artillery horses began to come back. One large horse, with his mouth shot off, clambered over our breastworks and fell down lengthwise in the ditch. Fortunately no one was caught under him, but he took up the room of eight men in the precious protection of the ditch. The horse was on his back and his legs were waving in the air. We rushed to get him out. Corporal Thomas bossed the job. We had no ropes or appliances, only the pieces of rail that had been broken by the round shot. We pried the horse up, swung him around and dragged him back of the ditch about ten feet. Then we were ordered to shoot every badly wounded horse that came back from the batteries in our front. We killed [a good] many.

All this time we dug ourselves in deeper. On our right, the batteries ceased fire, and we knew that this was to let the infantry pass beyond the guns. Then those batteries again opened up rapidly, giving the line of infantry all the protection they could by firing over their heads and delivering shrapnel among the Yanks. They took up the fire again, and we knew that our line had failed to take the position.

This was the first part of that charge which is called "Pickett's Charge." We had no watches and I cannot tell how much time elapsed between events. General Joe Davis' Brigade came up to our line from our left and filed along double-quick to our right just behind our breastworks. When they had gotten into position they were ordered to lie down. The Second Mississippi, of which Brother Joe was the surgeon, was immediately behind the Sixteenth Mississippi, and so he stopped to see if I was safe before he established his field hospital in a cut in the road not far to our rear.

We supposed that we would go as the first line with Davis's Brigade as our support. But as soon as that brigade got into position, they were ordered forward. They passed over us and ran on between the guns of the batteries which ceased firing to let them pass. They moved on into the open field and deployed, then advanced, and, as soon as the batteries could fire over them, they began a furious cannonade. The roar of the Federal guns constantly concentrated from both the right and the left upon the advancing line.

Then our batteries ceased firing and we knew that our line had advanced to the Federal position while the rattle of muskets told us that they were engaged in trying to storm the works. Again our batteries started up a violent shelling of the Federal positions, and informing us that Davis's Brigade had failed, and that our artillery protected a retreat. Soon the frazzled remnant of the brigade came staggering back through the smoke singly or in small groups. The organization of the brigade had been completely broken up.

General Davis, with his sword in hand, came slowly down the line. I knew him personally and went up to him and asked, "General Davis, where is your brigade?" He pointed the sword up to the skies, but did not say a word, and stood there for a moment knocking pebbles out of the path with the point of his sword. He could not talk, and neither could I. He walked on and I went back to my place.

Then came a lull in the firing. About one hour elapsed and then both sides opened up once more. Came the order from the right: "Form the line! Forward!" We went double-quick and filed between the guns which ceased firing. Then we reformed the line beyond the guns and advanced rapidly. The Yanks played on us with solid shot and shell. When we reached the small stream that ran at the foot of the hill and began the ascent of Cemetery Ridge, they changed to canister and grapeshot. We went, firing as he cold. We became enveloped in the smoke, though which we could see the fire of the guns. It looked as if we had charged into hell. Our line thinned down to mere groups of men, too few to accomplish anything. The order was given, "Fall back!"

We fell back over our dead and wounded, with here and there the body of a Yankee who had been left from the fighting of the day before. Our batteries did not fire to protect us and the Yanks had it all their own way. As we went back the guns we asked: "Why didn't you help us out?" And the dreadful answer was "Out of ammunition."

When I was a boy, the business part of the town of Woodville, Miss., burned down. The fire might have been checked but the water gave out. We stood around and saw houses burn which could have been saved if the town had made provision beforehand for an adequate water supply. And so much property was lost by the lack of foresight. The whole cause of the Confederacy had been lost by Lee's Army running out of ammunition.

"Pickett's Charge" of July 3, 1863, a fatal charge, named after the commanding major-general in the movement, failed and with it our hopes sank. Mississippians, Alabamians, North Carolinians, Tennesseans, and Virginians participated equally in this charge. There was nothing in it to brag about, except the heroic gallantry of the men. As far as the general officers were concerned, they could not have bungled up things more nor shown less military skill. Longstreet was the corps commander and was directly over all that part of the line.

He stopped us on the second of July when we had the Yanks going good, and when we were in sight of Cemetery Ridge, with reinforcements to continue the advance. He waited for more than twenty-four hours, which delay allowed the enemy to fortify and get in additional artillery and more troops.

When the charge was made, troops advanced by details. After each failure of a small force, the Yanks had ample time to prepare for the next small force. On four separate occasions this occurred.

The men argued many and varied questions. They could not comprehend the disaster which occurred, neither could they relish such methods leading to it. "What would General Lee do? Will he court martial Longstreet and, if so, what penalty will be inflicted?" was the typical of the questions asked. But "Marse Bob" took the blame on himself and let Longstreet off.

That night [July 3, 1863] it commenced to rain. Our skirmish line was ordered to advance just before day. The line was beyond the small stream and [we] dug rifle pits with two men to a pit. I was with Johnnie [in a] pit in a small depression that led down to the stream. We were well up on Cemetery [Seminary] Ridge. The rain increased and soon our pit began to fill with water. When day came the skirmishers became brisk. One of us had to bail out the water while the other shot. We took turns at bailing and shooting all day, soaking wet, hungry and tired [and] too busy to notice the flight of time. We had nothing to eat and could not have eaten if we had, without taking in quantities of that filthy water. We absorbed all the moisture that was necessary and then some.

* * *

WILEY PIERCE HEFLIN, BLIND MAN ON THE WARPATH (PHILADELPHIA, MS, 1905), P. 25.

PRIVATE WILEY P. HEFLIN
11TH MISSISSIPPI, COMPANY A

Fifty-five of my company went into that charge, ten reported to the company that night, only knew of five being captured, and not being wounded. The balance were killed and wounded. I suppose at least thirty-five of my company or two-thirds were lost in this battle. I was wounded on the left foot and ankle by a piece of shell and shot through my left hip with a Minnie ball. After our line fell back the Yankees advanced down the slope to where I was. They were capturing prisoners and passing them to the rear. I began to call to some of my own men to carry me off. They were all so excited they paid no attention to me. Then I began calling to the enemy to help me off, which I knew was against orders, but I had seen the wounded so much lie on the battle field for a day or two calling for water that I was after getting away from there. So one of the enemy helped me around on the opposite side of a little plank house near by, sat me down telling me then that I was out of danger. I hadn't been there but a few moments when a cannon ball from our own line came tearing through the house. Then I went to begging the soldiers in earnest to help me away. I told them to help me till I could get hold of a stick or something to support me, and I thought I could help myself. At last one agreed to do this and I got my arm around his neck and started, telling him to carry me a little further and a little further. I hung on to his neck until he carried me beyond the works to where the well prisoners were.

* * *

MAUD BROWN, THE UNIVERSITY GREYS, 33-34, 1940.

PRIVATE THOMAS FONDREN MCKIE, 11TH MISSISSIPPI COMPANY A

Camp near Winchester

Oct 29, 1862

Dear Mother,

I want you to write to President Davis to get me off; say to him that I joined contrary to your will and that I am a minor and that you desire me at home and are compelled to have me. Do it immediately as soon as you receive it. All the officers in our company advised me to do it. Do not hesitate to write because you half way consented for every one in the country knows that up to the very moment that I started you did not want me to come. You recollect on the very evening that I mustered into service, that we walked out into the garden and you cryed and asked me to go up to the depot and get off. But I fool-like did not want to do it. You had better write two letters for fear that one will be destroyed. My health is very good at present. My love to all my relatives and acquaintances. My love to all at home.

Goodbye

YOUR SON

[Ed. Note: Davis failed to excuse Private McKie and he fell mortally wounded during the Pickett-Pettigrew-Trimble Charge on July 3rd]

* * *

ALLEN J. HOOKER
DEMOCRATIC-HERALD (JACKSON, MS)

CAPTAIN JOHN R. PRINCE
11TH MISSISSIPPI INFANTRY, COMPANY D

Dr. J. R. Prince, Captain 11th Mississippi, in a letter to this writer, states that Lieut. Hester, who was captured at the stone wall, informed him that during his imprisonment a reputable Federal officer told him (Hester) that the dead of Mississippi were found higher up the Heights than those of any other command.

Dr. Prince wrote me that the himself gave the order to retreat on July 3, 1863, every commissioned officer having been shot down, and that the brigade retreated in passably good order, about 300 in number, back to its original position on Missionary [Seminary] Ridge, occupied by it in the morning. SO that if your correspondent will deduct the number of the 55th North Carolina (which he has) from the total, he can approximate the strength of the other three regiments just prior to the so-called "Pickett's charge."

* * *

PRIVATE O. R. GILLETTE
11TH MISSISSIPPI INFANTRY

SHREVEPORT JOURNAL
JULY 4, 1938

I accompanied Gen. O. R. Gillette of Shreveport and two newspapermen on a trip over the battlefield where General Gillette pointed out the spots where he fought at Gettysburg. He found a marker showing the position of Davis' Mississippi brigade, on the final day of the battle. It was just to the left of Pickett's main charge. Although the battlefield is greatly changed Gen. Gillette was able to point out the route he took during the charge. He went to Little Round Top and from this eminence where Meade's men poured a withering fire into the Confederate ranks he pointed in the spot where he and his comrades reached in the valley below.

A big crowd of spectators gathered around the general as he told the story of the battle. Children scrambled for his autograph. After Round Top, General Gillette went to Devil's Den, a mass of huge boulders where Confederate sharpshooters were stationed. Next he went to Lee's monument not far from Davis' brigade position. At the monument General Gillette met a jolly old Connecticut Yankee, who, among his decorations was wearing a badge of the Louisiana U. C. connection held at New Orleans 15 years ago. The old soldier was Ira R. Waldman, 88, of Danbury, Conn.

"Hello Johnny," Major Waldman called merrily to General Gillette. "Hello, Yank," cracked General Gillette. They met, shook hands, and engaged in merry conversation.

Major Waldman explained that a cousin of his was a member of a Connecticut force that captured a flag belonging to the famous Louisiana Tigers at Cold Harbor.

Gillette remembered too, how Meade's army of the Potomac reached out along the roads that centered like spokes of a wheel at Gettysburg and how they met there to fight and to forget they ever needed shoes.

"We opened the battle," he said, his strong hand sweeping out toward the fields west of town. "The town is different now as it was then, but there's two buildings I recognized right off. They say Lincoln stopped in one of them later to finish a speech he made here. I was in Davis' brigade, Heth's division, Hill's corps, the army of Northern Virginia.

There were others here today who remembered the battle too, one was Charles D. Clarke, of Austin, Texas who lay among the confederates watching the artillery a mile away across the valley.

Gillette the tall ram-rod backed southerner who came from his home at Bossier, La., remembered the last day most clearly of all. It was the final test - a thrust by Pickett's brigade at the Union center after attacks on both flanks who had failed.

"We were supporting Pickett's left," Gillette said. "There was heavy artillery fire from both sides. The field in front of us looked like ploughed ground where the shells hit."

"They went down like blackbirds," Gillette remembered. "When Pickett charged, we went too. The corn was knee high. We carried the flag. We went up the hill but we couldn't stay there. They've got markers now right on the spots where we went. It was a hot day and we fought.

* * *

SHREVEPORT JOURNAL
FEBRUARY 23, 1939
PRIVATE O. R. GILLETTE
11TH MISSISSIPPI INFANTRY

Gillette Given Missile Fired At Gettysburg Battle.

A minie ball, fired at the battle of Gettysburg over 75 years ago has been received by Gen. O. R. Gillette, commander of Leroy Stafford camp of Confederate Veterans who fought at the battle.

The minie ball was found on the battlefield recently by Lieut. Arthur Frey, stationed at Gettysburg. General Gillette met Lieut. Frey when he attended the Blue and Gray reunion at Gettysburg last July. At that time Lieut. Frey promised to send the general some memento of the battle.

"This ball," Lieutenant Frey wrote "was found by me in the ground taken from a post hole that was required for the erection of No. 3 mess hall. If you will recall this mess hall was in the Union camp area along the Carlisle road. You will also recall that this area was held by the eleventh corps (Union) the first day of the battle, that is, until they were forced to retreat through the town to Cemetery hill late in the afternoon.

General Gillette was a member of Davis' brigade, eleventh Mississippi regiment, which, with the eighteenth [second] Mississippi and Fifty-fifth North Carolina regiments, forced the retreat of the Union troops from this location.

The missile was identified as one of the type fired by the Confederates.

General Gillette gave to Dolph Frantz, editor of the *Journal*.

* * *

COLUMBUS WEEKLY DISPATCH
SEPTEMBER 14, 1905.

WILLIAM A. LOVE
GETTYSBURG RECORD

Col. Love Pays Tribute To Mississippi Troops In The Battle

Gen. Stephen D. Lee's recent review of incidents in the history of Co. E, 11th Reg. Miss. Inft. C. S. A., suggests the propriety, as well as timeliness of interpreting certain facts connected with the battle of Gettysburg, in which this company took a prominent part.

Someone has said: "To most of us history is a string of bright deeds, a sort of diamond necklace, the brilliance whose hems obscures the setting." So we see at Thermopylae only Leonidas and his three hundred Spartans; from Balaclava we hear only the thundering hoofs of the Light Brigade; at Waterloo we see only Ney and the Old Guard, with back to the field and face to the foe, and at Gettysburg only Pickett and his Virginians marching on to glory.

"Steadily they step down the slope,
Steadily they climb the hill.
Steady they load, steady they fire,
Marching right onward still."

However obscured forty years ago by the froth of poetic fancy and the ebullition of crude historians, post bellum literature in the last instance at least has relieved the setting of much of its obscurity.

To a student of Confederate history a bare mention of facts leading up to the action of the third day at Gettysburg in unnecessary. At the time the army of Gen. Meade was posting behind the stone wall crowning Cemetery Ridge. This wall is an irregular line north and south. Near the center, it recedes thirty yards, forming what is known as the "Angle." Three-quarters of a mile to the West lay the army of Gen. Lee. An assault of the Federal position having been determined upon, to Gen. Longstreet was assigned the duty of its direction. Of the nine divisions present, two, those of Genls. Heath and Pickett, were selected as the assaulting columns under command of the latter. After the most terrific and prolonged artillery duel in the annals of American warfare the column advanced, Pickett's division on the right, Heath's on the left. A part of Armstead's brigade of Pickett's crossed the wall near to and South of the point of the Angle, Gen. Armstead falling thirty three yards within the enemies lines. The objective point of Heth's division being to the North or left of the Angle the distance traversed was necessarily thirty yards greater. A part of the division, however, reached the wall and just here the "high watermark of the Confederacy was reached."

Why a part of your division was able to cross the wall while the other failed to do so, and which "went farthest to the front,' are subjects of discussion by surviving participants of Virginia and North Carolina. Tennesseans and Mississippians modestly withhold special claims, preferring to leave to correctly informed posterity the privilege of deciding what a soldier must endure what obstacles he must overcome and how far to the front he must go to constitute himself a hero. As an aid in that direction, simile for the present is discarded and statements of participants on both sides, Federal and Confederate, are collated.

First Federal Gen. Doubleday states that all the guns with a single exception in front of Pickett's division were silenced by the Confederate batteries. The author of "New York at Gettysburg," (42d N. Y. Tammany) says: "Longstreet's assault on the third day had some slight resemblance of success when Armistead and his men crossed the wall. "Had there been a strong force of infantry on that portion of the line, not a Confederate would have crossed the wall."

Capt. Wood, of Armistead's brigade, gives this account of the charge: "From the time the charge began up to this moment not a shot had been fired at us, nor had we been able to see because of the density of smoke which hung over the battlefield like a pall, that there was an enemy in front of us. The smoke now lifted from our front and there right before us, scarcely two hundred yards away stood Cemetery Hill in awful grandeur."

Gen. Hunt, commanding the Federal artillery states: "I had counted on an artillery cross-fire that would stop Pickett's division before reached our line but except a few shots here and there, Hazard's batteries were silent until the enemy came within canister range."

Second. Maj. Hall, of Fifth New Hampshire Regiment, says, "There is not a shadow of a doubt in my mind but that the sons of North Carolina, Tennessee and Mississippi carved on the tablets of history equal laurels with the sons of Virginia on the tablets of history equal laurels with the sons of Virginia on the great event of that supreme attempt to gain victory at Cemetery Ridge. Maj. Hall was not only a participant in the battle, but the proprietor of a Cyclorama representing its scenes."

Maj. Robbins, of Fourth Alabama Regiment and Gettysburg Battlefield Commissioner, said: "And while Armistead and his heroic followers were over in the Angle, where were Pettigrew's thin but gallant batteries? They were making a desperate effort to storm the high stone wall eighty yards east of the Angle and were being mowed down like gran before the reaper by the double line of infantry behind the wall. A few men did reach it, but finding it too high to leap over could do nothing but surrender."

Col Bachelder, Federal Battlefield Commissioner, wrote: "The left of the column continued to move on towards the second wall, threatening the right and rear of Gibbon's division which held the advance line, Gen. Webb, whose brigade was on the right (in the projection) had hurried back to bring up his right reserve from the second line. But before this could be accomplished the first line broke under the terrible pressure which threatened its front and flank fell back upon the reserve."

Gen. Hancock sent Gen. Meade this dispatch: "I had to break the line to attack the enemy in flank of my right where the enemy was most persistent after the front attack was repelled."

Excerpt from Committee Report N. C. Historical Society 1904: "It is well known that the left of the line was subjected to a severe artillery fire from almost the very start. Many of the regiments suffered greatly before the right had lost a man by shot or shell. In this assault the left brigade which had done indifferent fighting from the first day, now did none at all. The enemy witnessing their conduct naturally concluded that the left was the weak spot and acted vigorously upon that conclusion. Davis' brigade which came next, the Virginia which had acted so ingloriously, was composed of the 2d, 11th and 42nd Mississippi and 55th North Carolina. Three of these regiments had fought the first day, had gained a victory but with great loss. The 11th Mississippi was on detached duty that day and though it and its brigade did not go quite as far in the assault as troops on their right the loss in killed and wounded for the regiment was 202, which was over 60 per cent. Pickett's men had 91 killed and wounded to the regiment, or about 28 per cent. When all things are considered, their bloody fight on the first day, their standing up so long and so manfully this day against so terrific a fire both from flank and front (as sown by their losses) it will be seen that it is possible that it is possible that "Truth" may proclaim that Davis' brigade, though they did not reach the vicinity of their works, are more deserving of the palm than any other command on the field. For it is not known positively that any other brigade lost on that day so large a percent of killed as they."

Virginians were quietly chagrined at having to abandon works once entered. North Carolinians, Tennesseans and Mississippians were equally mortified at their inability to enter; and so where all did their duty, there should be no criminations and recriminations for are they not all Americans.

* * *

Colonel John M. Stone
2nd Mississippi Infantry

Clarion-Ledger
April 27, 1892

The Second Mississippi Send a Piece of Their Flag, With History, to Gov. J. M. Stone

A Clarion reporter was shown a ragged, blood-stained fragment of a battle flag yesterday, that was sent to Gov. Stone, with the accompanying letter, by Dr. Joseph Holt, surgeon of the Second Mississippi Regiment. The little scrap of bunting must be woven through with tender recollections; for the tears rose to the eyes of more than one strong, brave man, as the hollowed souvenir was passed yesterday from hand to hand.

New Orleans, La., April 16, 1892.
His Excellency Gov. J. M. Stone.

Sir – Enclosed, I have the honor to transmit to you a portion of the regimental battle flag of the Second Mississippi, which flag was not surrendered, but brought home from Appomattox Court House, secreted on the person of "Billy" Burns, the flag bearer, who subsequently removed, with many of the others of the old regiment to

Texas, from whence six of them (not including Burns) whose names and addresses are hereto appended, came to participate in the "Reunion of Confederate Veterans," recent held in this city.

On that grand occasion, which was indeed most impressive and a great success, the Second Mississippi contingent, marching under the tattered battle flag which was proudly held aloft by Private John H. Allen, (who lost his left forearm to the cause) was a center of enthusiastic greeting among the whole line of march. The position assigned us was at the head of the divisions, purposely selected as conspicuously befitting the war-worn colors and the one-armed bearer.

A (to us) memorable episode of the occasion was the warm reception tendered the guard of honor to the flag, by Senator Daniels, who expressed an earnest desire to make the acceptance of the Second Mississippians, and to whom I introduced them at the residence of Gen Behan. The profoundest expressions of admiration and of honor were bestowed. Private Allen being especially distinguished, although Mr. Gray, the gallant fellow, deserved quite as much, in consideration, of a deep groove he carries in his skull, ploughed there by a rifle ball.

I have just received from Mr. Gray a large and very handsome photograph of the flag, with a group of the "boys" now living in Texas.

I hope and have reason to believe that you have already received a like copy. We join in presenting you these mementoes, feeling assured you will accept them as souvenirs of your own splendid war-history, into which ours is so intimately woven.

Please accept this as a conjoint letter to which through the authorized writer, we subscribe our names, with assurance of exalted esteem for your personally, and of the utmost pride in you as our regimental commander.

Wm. Burns, Company B., Blossom Prairie, Texas; John H. Allen, Company B., Taylor, Hood county, Texas, Isaac N. Gray, Company B., Cooper, Delta county, Texas; John A. Mayfield, Company C, Itaska, Hill county, Texas; L. M. Champion, Company D.; Jerome B. Patton, Company F., Stevenville, Erath county, Texas, J. R. Bolin, Company H.; Joseph Holt, M. D. Surgeon.

Dr. Holt's letter is inaccurate in this:

The flag bearers name is "Byrn, not Burns, and the flag was never at Appomattox Court House."

Billy Byrn, the flag bearer, was shot down in the battle of Gettysburg, and being unable to walk was left on the field. He took the flag off the staff, threw the staff away and concealed the flag in his clothes on his person. He was taken to the Federal hospital, thence to prison at Fort Delaware, where he remained until the close of the war, and when discharged from prison he returned to his home in Tippah county with his flag, having kept it concealed during his imprisonment, and his arrival at home.

J. M. Stone,
Jackson, Miss., April 23, 1892.

[Dr. Joseph Holt was the surgeon of the Second Mississippi Regiment.]

* * *

VICKSBURG EVENING POST
MAY 3, 1902

Mississippi Flag Found Was Hanging Up in a Saloon at Dallas.

Hon. C. B. Curlee, who attended the reunion at Dallas, writes to *Corinthian* this interesting bit of news:

"We have something to relate that will gladden the hears of all the Second Mississippi Regiment. Nat Bynum came into Mississippi headquarters this morning and told us that he had seen the old battle flag of the Second Mississippi Regiment hanging up in a saloon in this city. A posse was immediately made up of Major G. W. Bynum, Turner Bynum, John Burge, Billy Kimmens and your correspondent to get the flag and restore it to its own. I told the glad news to Rev. J. E. Gambrell and invited him to go with us "Where is?" he asked, I replied: "hanging up in a saloon down on Camp Street." He gathered himself together as if he had received a blow in the breast, his smile faded from his face and I knew his thoughts had reverted back to years ago when he passed through the greatest trial of his life. We found and secured it at last. It was written upon its folds, '2nd Mississippi Reg.' 'Manassas,' 'Seven Pines,' 'Gaines Farm,' 'Malvern Hill,' '2d Manassas,' 'Boonesboro,' 'Sharpsburg,' 'Gettysburg.'

"At the surrender Nat Bynum tore it from its staff, hid it in his bosom and it was never surrendered. The flag has been lost for thirty-seven years and we are all rejoicing that it has been found and restored to the possession of the old vets who fought under its folds."

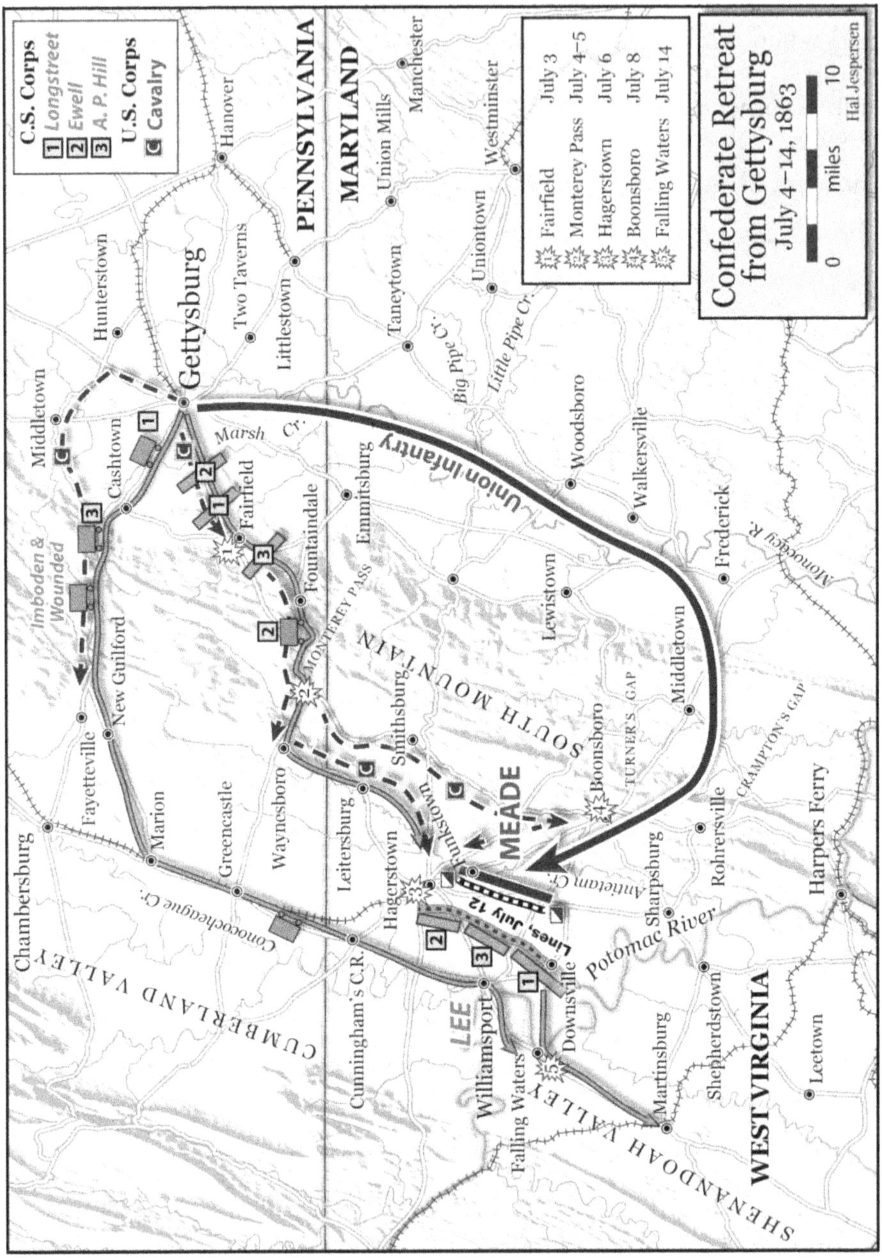
Confederate Retreat
from Gettysburg
July 4–14, 1863
0 miles 10
Hal Jespersen
C.S. Corps
1 Longstreet
2 Ewell
3 A. P. Hill
U.S. Corps
Cavalry
1 Fairfield July 3
2 Monterey Pass July 4–5
3 Hagerstown July 6
4 Boonsboro July 8
5 Falling Waters July 14
PENNSYLVANIA
MARYLAND
WEST VIRGINIA
CUMBERLAND VALLEY
SHENANDOAH VALLEY
SOUTH MOUNTAIN
Gettysburg
Union Infantry
Imboden & Wounded
LEE
MEADE
Lines, July 12
Potomac River
Chambersburg
Fayetteville
New Guilford
Marion
Greencastle
Waynesboro
Leitersburg
Hagerstown
Cunningham's C.R.
Williamsport
Falling Waters
Downsville
Martinsburg
Shepherdstown
Leetown
Harpers Ferry
Sharpsburg
Rohrersville
Boonsboro
Middletown
Lewistown
Frederick
Walkersville
Woodsboro
Emmitsburg
Fountaindale
Fairfield
Cashtown
Middletown
Hunterstown
Two Taverns
Littlestown
Hanover
Union Mills
Manchester
Westminster
Uniontown
Taneytown
Smithsburg
Funkstown
Monterey Pass
Turner's Gap
Crampton's Gap
Marsh Cr.
Big Pipe Cr.
Little Pipe Cr.
Monocacy R.
Antietam Cr.
Conococheague Cr.

CHAPTER FIVE

Retreat From Gettysburg and Post War

As the gunfire subsided on July 3rd Lee and his battered army prepared for a possible Union counterattack as best they could -- an assault which never came. Having been repulsed at Gettysburg, Lee realized he could not remain in a vulnerable position there in enemy territory with no hope of a decisive victory and that night began preparations for a withdrawal. The Grey Fox contracted and straightened his defensive line by withdrawing Ewell's Corps from Culp's Hill back through Gettysburg and positioning it along Oak and Seminary Ridges. Overnight his Confederates built breastworks and dug rifle pits which extended in a line two-and-one-half miles long from the Mummasburg Road to the Emmittsburg Road. All during the day of July 4th the Army of Northern Virginia remained nervously behind these formidable defenses inviting a costly Federal attack. But General Meade remained content to merely observe his enemy and lick his wounds. That night the Confederate retreat commenced with Hill's Third Corps taking the lead along the Fairfield Road, soon followed by Longstreet's First Corps, and finally by Ewell's Second Corps. Screened by Stuart's cavalry this procession of infantry and artillery proceeded through Monterey Pass headed for Hagerstown, Maryland. Late in the afternoon the immense Confederate wagon train laden with captured supplies and ambulances carrying the wounded, commanded by Major General John Imboden and guarded by 2,100 mounted men and a battery of artillery, had departed westward along the Cashtown Road. Proceeding through Cashtown Gap they headed for Hagerstown via Greenwood and then Greencastle.[1]

Contrary to claims by the Northern press and later testimony presented to the Committee on the Conduct of the War by Union officers hostile to Meade, Lee's army was not "a demoralized horde of fugitives". Though exhausted and badly mauled at Gettysburg, these Confederate veterans remained a potent fighting force, confident and capable of striking a crippling blow to a careless adversary. Indeed soldier diaries and correspondence bear witness to this truth. Typical is an extract from a letter by a Mississippi officer in Posey's brigade dated July 19, 1863: "The truth is we suffered a severe repulse at Gettysburg,

1. Coddington, *The Gettysburg Campaign*, 535-38.

(although we inflicted a greater loss upon the enemy than we ourselves sustained) *but as to our army being utterly destroyed and demoralized, it is sheer nonsense.* (emphasis added) The Army of Northern Virginia cannot be routed." His veterans still believed in their commander, and Lee still remained hopeful of striking a decisive blow against his enemy should the opportunity present itself.[2]

To add to their misery a downpour began during the evening of July 3rd and continued throughout the next day. In fact it rained almost continuously for the next week turning the roadways into a muddy quagmire that took a mighty toll on their dilapidated footwear which rendered many a foot soldier barefoot. Lack of available provisions and only meager spoils from foraging efforts added extreme hunger to their suffering. And, despite the absence of Union infantry, there was the ever-present fear of being overtaken and captured by Union cavalry patrols which quickly spread over the countryside.

The wagon train proceeded from Greencastle directly to Williamsport to ford the Potomac there while Imboden returned to Hagerstown to guard the army's route to Williamsport and Falling Waters. By July 11th Lee had selected a site for his defensive line here along high ground and threw up formidable earthworks while his engineers constructed a pontoon bridge across the Potomac River at Falling Waters. Then Lee waited and hoped for a Union attack which was ultimately scheduled by General Meade for July14th. With the bridge completed and the Potomac River now fordable, Lee ordered his army to cross over into Virginia during the night of July 13th. Hill's and Longstreet's Corps with the artillery crossed on the pontoon bridge at Falling Waters while Ewell's Corps waded across the Potomac at Williamsport. Henry Heth's Division, together with Pender's, formed the rear guard at Falling Waters and were surprised at dawn by a Union attack which netted 719 prisoners.

As Edwin Coddington noted: "Ironically, Henry Heth, who had opened hostilities at Gettysburg, also fired the closing volleys fourteen days later at Falling Waters -- and in both instances he was caught unprepared." Once again Mississippians were "in the very thickest of the fray." After suffering 341 casualties at Gettysburg eleven days before, the 11th Mississippi lost thirty-six more in this brief rearguard action. Indeed, Stephen Stubbs, chronicler of the 11th Mississippi, penned this stirring tribute:

> **After crossing the Potomac and scrambling up the bank, Private Gabe Smither, of Company G of the 11th Mississippi, in passing the regimental band, said to the**

2. Ibid, 536; William Cowper Collection, Archives and Special Collections, J.D. Williams Library, The University of Mississippi.

Sixteen Survivors of "Lamar Rifles" at Reunion at Bem Price's, Oxford, Mississippi, November 28th, 1901—Thanksgiving Day.

(LAMAR RIFLES: A History of Company G Eleventh Mississippi Regiment, C.S.A.)

> **leader, "Stewart, by blood, play Dixie." Soon the quick notes of that ever inspiring air "wafted upon the breeze." Then followed a roll of the "Rebel Yell" of defiance that meant too plainly to the enemy on the other side of the river that there was yet remaining strength, determination, and fight in the Army of Northern Virginia. To Heth's division, of which the veteran soldiers of Mississippi had formed an integral part, belonged the honor of fighting first at Gettysburg and last at Falling Waters on the Potomac River.**[3]

Safely back on Virginia soil, Surgeon Legrand J. Wilson of the 42nd Mississippi lamented the severe losses suffered by these Mississippians during the Gettysburg campaign with these touching words: "When I saw my regiment on dress parade for the first time after our return to Virginia, the change the short campaign had made caused a pang of sorrow. The regiment was in command of a Captain. The Colonel dead, the Lieutenant-Colonel, a hopeless cripple and the

3. Coddington, *The Gettysburg Campaign*, 571-72; Stubbs, *Duty*Honor*Valor*, 448.

Major severely wounded. Our gallant Adjutant escaped. Many of the company officers gone *and in the ranks, great gaps, great gaps!*" (emphasis added)[4]

* * *

3D DAY AT GETTYSBURG
COPY IN 17TH MISSISSIPPI FOLDER, GNMP

PRIVATE JOHN S. HENLEY
17TH MS, COMPANY A

[The Retreat From Gettysburg]

We commenced moving back around dark, and marched all night and all the next day until about four o'clock in the evening, when we reached the Potomac River. I had not slept at all for 48 hours. A pontoon bridge was thrown across the river near Winchester, and we laid down and slept for about 20 hours.

* * *

VICKSBURG EVENING POST (VICKSBURG, MS)
1896

SGT. CHARLES E. HUTCHINSON
CO. H, 48TH MISSISSIPPI, CO. H

Soldier Diary

July 4 – Heavy skirmishing today all along the lines. We were moved further to the right, and again threw up temporary rifle pits with bayonets and tin plates. Heavy rain most of the day. We will fall back tonight, and it will fall to our battalion to go on picket tonight.

July 5 – We relieved the 16th [Mississippi] pickets last evening and the artillery were withdrawn, and in an hour, all the troops in our front were gone. I did not like to be left but the night passed very quietly, with occasional picket firing. Rained hard all night. At the break of day, the order was given to fall back, and we went at a double quick to the rear and formed. Here had the hardest march I remember. Passing through Fairfield we soon struck the mountains 8 miles from Barnett's Ford. Raining all the time and the roads knee deep in mud. After we got to the mountains, the narrow road was blocked with wagons, and such a march [I] have never seen. Our brigade was ahead of us, and we were trying to get up to it. We passed over and through the mountains and camped at night 1 mile from the Gap, and near a

4. Wilson, *The Confederate Soldier*, 136.

little town – Waterloo. Our wagons were not up, and we will have to forage some. We have marched between 20 and 22 miles. The enemy have made some captures from us in the mountains.

July 6 – Lay in camp until all the army had passed, and we were rear guard. marched all night. A private in the 12th was shot and killed by Sergeant Gregory of the 19th. He was drunk and will die for it. Marched until 8 o'clock and stopped to cook. All as hungry as wolves.

July 7 – In camp near Hagerstown cooking.

July 8 – Still in camp near Hagerstown. Rained very hard last night. Our tent fell down and we got very wet. Still raining. There have been but two or three days it did not rain since we crossed the Potomac.

July 9 – Still in camp. Today has been beautiful. Have heard all kinds of rumors today. Heard "that we had possession of New Orleans," "that Grant is surrounded and his supplies cut off," that "the enemy lost forty-two thousand men at Gettysburg," and that "we have possession of Maryland Heights." God grant it may all be true. The enemy came near getting our wagons today, but was driven off by our cavalry. I am in hopes that peace may soon be declared, and that we may be sent to our homes all right. Wise counsel must prevail; fighting will not make peace. Sent letter home today, and think it will go through this time.

July 10 – Pleasant day. Fighting in the direction of Franktown; we moved at 3 o'clock. Sent out foraging party for battalion. King went from our company. We passed through Hagerstown today. Have changed my opinion in regard to the place. There are a great many Southerners there, and some very pretty girls. We are now camped, I suppose near Franktown. Expect a fight tomorrow. "God defend the right."

* * *

WILLIAM H. HILL, DIARY 1861-63, 2 VOLS.
Z0217.00
MISSISSIPPI DEPARTMENT OF ARCHIVES AND HISTORY

WILLIAM H. HILL
SGT. 13TH MS, COMPANY H

Diary Entries

July 4 - Saturday. Morning clear, heavy rain in the evening and continued all day. We left camp at 4PM and started in a southwest direction. General Hill's Corp in advance. General Longstreet's next and General Ewell bringing up the rear with the 2nd Corp. We left all of our wounded, who could not walk in the enemies hands and left nurses to attend them. Brig. Gen. Semmes was severely wounded in the leg. The femoral artery was cut. Maj. Gen. Hood was wounded in the arm.

July 5 - Raining all day. Left camp at 4 am, passed through Fairfield, crossed the Blue Ridge Mountains at Fountdale gap. Marched all day and night. The enemy pursued us a short distance this morning but Gen. Ewell defeated them so badly that they stopped the chase.

July 6 - Monday. Cloudy and misting rain. Marched all night, stopped at 4am at Montey [Monterey] Springs on top of South mountain. Cooked breakfast and resumed the march at 6 am. Marched yesterday and last night 18 miles. Roads are very bad. Passed through Ridgefield and Leitersburg, arrived at Hagerstown at dark. Stopped there for several hours and resumed the march to camp 1-1/2 miles south of town. The Cavalry of the two armies had a severe fight at Hagerstown today. Our force defeated them with severe loss.

July 7 - 8 missing

July 9 - Thursday. Cloudy and rain. The wagon trains moved to Downsville where the brigade is on picket.

July 10 - Friday, Clear and warm. It is now ascertained beyond a doubt that Capt. Robert Topp and the detail of 10 men that were with him were captured on the 4th inst. Nine belonging to our Regiment and one to Headquarters. The enemy also captured that same day 2 wagons that were sent out after forage.

July 11 - Saturday. Clear in the morning, balance of the day, cloudy. The brigade moved 2 miles north toward Hagerstown and threw up fortifications. Our pickets are skirmishing with the enemy.

July 12 - Sunday. Cloudy and raining all evening. The skirmishing still continues.

July 13 - Monday. Heavy rain last night, raining again today. Our wagon trains commenced crossing the Potomac at Williamsport, Maryland this morning. The river was very high and several wagons, teams, ambulances, and men were lost in the crossing. Marched to Falling waters and camped. Distance marched 6 miles.

July 14 - Tuesday. Cloudy and raining all evening. All our army left the trenches last night and fell back across the river on the pontoon bridges at falling waters. The enemy are in pursuit. There was a severe skirmish at falling Waters this evening. Marched 14 miles and camped 3 miles from Martinsburg.

* * *

Pvt. David Holt
Civil War Memoir, 200-204
Original Typescript Manuscript at MDAH

Private David E. Holt
16th Mississippi Infantry

Several times our pit was struck. Once a ball struck the mud just opposite my head, the mud splashing into my face. I dug out the bullet and it had flattened out to about the size of a half-dollar. We lost some additional men on that skirmish line. We remembered that it was the Glorious Fourth [July 4th] and we were having fireworks. At last night came again, and we were withdrawn.

We never halted at our old line of breastworks but continued on a night march. The rain fell on the road in about the right proportions, and the tramping troops acted as a sort of mud mill, working the mud into the right consistency to make brick. It clung to our feet and to our clothes. When one stepped into a rut, he found it a hard matter to pull out his feet.

My shoes were of Confederate half-tanned leather, the soles being pegged on. They had a preliminary soaking all day in that skirmish pit, making them well-prepared for the trick that they contemplated playing on me. It was so dark that it was impossible to pick the way. I stepped into a rut that was nearly knee-deep. I pulled out one foot, and the sole of that shoe came off. Then I pulled out the other, and the sole of that shoe came off also. There I stood, shoed but shoeless, with uppers, so that I carried over ten pounds of mud per foot. I reached down in the dark and cut the strings and scattered the uppers, like the last rose of summer, fondly over the road. My socks held out for a while, then the bottoms wore out while the mud began packing around my legs inside of the socks, as if to take a plaster cast of them.

We marched rapidly, and I had no time to adjust things, so I pulled off the socks and threw them away. Now I had my feet on the ground, and they surely found all the jagged stones within reach. I got out of the road and walked along the side, on grass. I was getting along first rate, when the brigadier general came along and ordered me to get back into the road. I said I was bare-footed and that the stones cut my feet. He stormed out that he was not going to allow a private to talk back to him, and he called the guard with orders to take me in charge and bring me to headquarters at daylight. The general rode on, and the guard grabbed me. I do not know who the sergeant was, but he said, "Are you David Holt? I said that I was. "That general is a fool to expect a barefoot man to walk in that road," he said. "Walk where you can."

...We halted by a small stream that was swollen and muddy, and waded in and washed the mud off our clothes...

We went over South Mountain, and while in the mountains, one of the men noticed a tree of mountain birch. He said the bark was sweet and tasted like wintergreen. He stepped out of ranks and brought back a sample. Soon every man in the company was chewing bark. We had to get it as near to being powder as possible in order to eat it. That was all the food we had for the fourth and fifth of July. There were very few blackberries along that road.

In the defiles of South Mountain we came upon some of our army wagons that had been burned by the Federal cavalry in a bold dash, but Jeb Stuart, with his cavalry drove them off. We knew that Jeb was defending our flanks. We gathered wheat, which was in "the milk," rubbed off the chaff with our hands, and boiled it in our little cans at night. It never got soft by boiling, but it was food, and we had the digestion of ostriches.

The only good filling we got from Gettysburg to the Rapidan River occurred when we halted near William and Mary College. That famous old school had furnished intellectual pabulum to Winfield Scott and other near-great men and now [it] furnished material nourishment to the Sixteenth Mississippi. Starving men prowl around like the "roar lion seeking what it may devour." Some of the Sixteenth Mississippi went over to the college to see what they could find. They found a building locked up and as quiet as a tomb. One of the investigators noticed in the basement area a solid wooden door and beside it a grated opening too high to reach from the ground. He began smelling around it like a hound on a cold trail.

"Boys, this is a storeroom," he said, "and I smell bacon in it. One of you let me stand on your shoulders, and I can reach that opening and look in." [We] raised him up and when he looked in, he said, "By the holy smoke, but the place is cram full of rations." They managed to break in and come back to camp loaded down with bacon and sacks of flour.

Colonel Baker had them bring it all to the regimental quartermaster and had the word passed along from company to company, "That any man was privileged to go and get all the rations he wanted." Then he stationed a guard around the camp and had them take everything the men brought. Then it was assembled and issued as rations. We all got full for once.

[But that came later. Back on the march,] we moved along leisurely. The weather was good, the road improving and our spirits picked up.

We lacked rations and some of us lacked shoes, but there was no grumbling.

* * *

Brookehaven Semi-Weekly Leader
October 7, 1908

Private Beverly Young
11th Mississippi Infantry

Southern Heroism

Beverly Young - A Pathetic War Story by Dr. B. F. Ward.

Beverly Young was a private soldier of "Co. I," Van Dorn Rifles," 11th Mississippi Regiment. He was one of the first to enlist and served as a private till he received his fatal wound in the desperate charge at Gettysburg. He was one of the noblest men I have ever known.

Though wealthy, he was modest, assuming and unpretentious yet genial, companionable and courageous in the face of every danger and under the burden of any hardship. He was severely wounded in the leg in the last day's battle, but with a crooked limb which he had found and used as a crutch, he could hobble around a little, and insisted on my giving attention to others whom he considered, were in more urgent need than himself, saying he could wait. For the benefit of protection, from a July sun, I had laid a number of very badly wounded on the ground in an Oak grove, through which ran an insignificant little brook that you might step across. The night after Gen. Lee retired, leaving five or six thousand wounded on the field, an unexpected rain fell in great and rapid torrents. In a few minutes the little stream had overflowed its banks and flooded the ground where my wounded lay, from ankle to knee deep. It was very dark and we had nothing but candles with which to make lights and they could not be used in the rain. We were apprehensive that some of the brave fellows would drown before we could get them out, but we finally found and rescued them all. I waded in water knee deep to where I knew one man was lying with his head at the root of a large oak. Hearing nothing from him, I thought he was drowned, and stooping down to feel from him. I touched an uplifted hand wet and cold as the chill of death. I said are you dead or alive? He recognized my voice and faintly answered, "Doctor, I thought I was gone and was holding up my hand in prayer." I said, your prayer is answered for the present, at least. The water was over him up to his chest, but he had pulled himself up till his head rested against the root of the tree sufficiently high to keep above water. I took him by the arms, pulled him into a sitting position and kneeling down in the water with my back to his chest and making him clasp his arms around my neck, I staggered to my feet and carried him to high land, although his weight was greater than mine.

When daylight came we were confronted with a said spectacle. The poor fellows were wet and chilled, their wounds were soaked with dirty water and none of them had a change of clothes. Not far away was a farm house, and a very large brick barn, which some of the boys informed me they thought was filled with straw. I walked up to the house and was met by a typical Pennsylvania farmer. I saw the hard lines in his face and realized the rigor of the Judgment Seat before I stood. How I did wish, that morning, for the eloquence of Paul that I might have made that old Hessian Felix tremble. I did my pathetic best, but could not reach him. I wanted straw worse than the Israelites did in the brickyards of Egypt, but could not have it without the money. I told him that I had not a dollar of funds, that would be current in his country and reminded him that, on the great battlefields of Virginia, many Federal wounded had fallen into my hands and that I had always extended to them the same care and attention given to my own wounded, but he was not in the "reciprocity" business. I then inquired how much he would charge for straw to make beds for a certain number of wounded, as I only wanted it for the worst and most helpless cases. He replied that I might have the straw for five dollars.

I walked back with a heavy heart and nave never liked to even write the name of Pennsylvania from that day to this.

I feel, still yet, the anguish that seized me when I looked on the pale and pinched features of those boys a thousand miles from home dying upon the bare, wet, inhospitable soil of a heartless enemy, without even an old Confederate blanket between them and a dying couch. I told them I had not a dollar with which to purchase the straw to make them a bed. Bev. Young was sitting down by a tree waiting for his clothes to dry on him. He put his hand in his pocket, took out a five dollar gold piece, the last he had, and that saved through many months of hunger and want for such an emergency as he had now overtaken him, and said, "here Doctor, take this and pay the old Shylock for his straw." I said, "Bev, you are going from here to a Yankee prison; you are badly wounded and will sorely need that money. I do not feel it is right to take it from you." He said, "Yes, take it and provide for those boys, they are worse off than I am. I went back, dropped the coin in the old man's hand and his eyes glistened like a basilisks as he clutched it. I think his greed was the more gratified because it was blood money.

We went to work packing the straw in our arms, stripping the boys and covering them, naked, in the stray while we hung their old soiled and bloody garments out to dry. Brave, noble, unselfish, great-hearted Bev Young crippled, penniless to the cruelties and inhumanities of a Northern prison and died of wounds. The people of Columbus ought to build a monument to his memory with a five dollar gold coin embossed upon its face and an inscription under it something like this, "Beverly Young, wounded and a prisoner, gave his last dollar to his suffering and dying comrades."

* * *

WILLIAM COWPER COLLECTION
ARCHIVES AND SPECIAL COLLECTIONS
J.D. WILLIAMS LIBRARY
THE UNIVERSITY OF MISSISSIPPI

WILLIAM COWPER NELSON
ORDNANCE OFFICER CARNOT POSEY'S BRIGADE
(FORMERLY 17TH MISSISSIPPI INFANTRY)

Letter to Maria C. Nelson,
Head Quarters Posey's Brigade, Near Bunker Hill, Va.
July 19th, 1863

Dear Mother

Although it is with faint hope of the letter reaching you I take advantage of a day's rest, to write you, assuring you of my continued good health, and safe arrival back into Virginia after a short visit to the United States: as I predicted in my last, we did actually invade the enemy's territory, and went into Pennsylvania...

From Chambersburg we marched to Gettysburg where was fought the bloodiest and most terrible battle of the War. God grant that it may be the last one. The battle commenced on the 1st of July. Two divisions, Pender's and Heath's of the 3rd Corps were engaged on that day, they drove the enemy several miles that day. The next day the remainder of our army having reached the scene of the action the great battle began in earnest. Longstreet came around on the right and endeavored to turn their left, this he partially succeeded in doing, driving them before him for more than a mile, but as one crest of hills succeeded another, each commanded by the succeeding one, it was impossible to carry the heights. And he was forced to fall back. Barksdale's Brigade sustained a terrible part in this engagement. The brigade lost over seven hundred men. General Barksdale was killed and almost every Field officer was either killed or wounded. My old company in the 17th lost 29 killed and wounded. Franklin Ross and

Willie McRaven of the Rangers were the only ones of your acquaintances who were killed. David McWilliams had a leg amputated and was left behind on our retreat. Frank Smith, Jimmy Crump, Stagey Watson were all unhurt. Col. Holder was severely wounded in the face. Jerry Gage of the University Greys was killed, he was a great friend of Miss Janie Edmondson's. Please mention the fact of his death to her.

In the 3rd another unsuccessful attempt was made to storm the heights and on the night of the 4th General Lee fell back to Hagerstown where we remained a week waiting for the Yankees to follow up their victory, but although they say that the rebel Army is routed and demoralized they did not see fit to attack us: and on the night of the 13th of July, 10 days after the fight, we safely crossed the Potomac. The truth is we suffered a severe repulse at Gettysburg, (although we inflicted a greater loss upon the enemy than we ourselves sustained) but as to our army being utterly destroyed and demoralized, it is sheer nonsense. The Army of Northern Virginia cannot be routed. Had it been we could never have crossed the river in safety, in the face of an overpowering force...

* * *

FORT WORTH DAILY-GAZETTE
MAY 31, 1887

C. C. CUMMINGS
17TH MISSISSIPPI INFANTRY

Evening Services

The evening services were held in the opera house which, in spite of the intense heat, was filled to overflowing by the youth and beauty of our city who came to grace the occasion and by the elderly who came to join in rendering homage to the memory of the departed dead. The stage was becomingly arranged. A large cenotaph stood near the footlights over and around which was folded an American flag. This was in memory of the unknown dead who sleep in unmarked graves. A little back and to the right hung the picture of President Lincoln; beneath and to the right was a picture of General Grant, and underneath it a picture of Garfield. Directly underneath was a picture of President Arthur and to the right was that of Lee and a large steel engraving of Lee and Jackson. In the center of these pictures was a map showing the position of the troops at the Battle of Gettysburg on July 3, 1863, at 3:10 p. m. The map was planned by Judge Cummings in a few preliminary remarks. Above the map was the picture of two soldiers, one in blue and one in grey, lying, as thousands have lain, cold

Sergeant Major Charles C. Cummings of the 17th Miss. Infantry

(Fort Worth Star-Telegram)

and dead upon the battlefield. The map was of great assistance to the audience enabling them to fully appreciate the situation. In front of the stage the Post was seated: Confederate veterans were seated at the right front of the stage; Texas and Mexican War veterans were assigned the seats of honor in the boxes; Commander Emerson, Adjutant Brown, the representatives of the press, the vocal music corps Elder Gavin and his reverend gentlemen companions, His Honor the Mayor, officers of the city government and members of the City Council, the Hon. Judge Cummings, and a volunteer platoon composed of Texas, Mexican, National and Confederate war veterans, were seated on the stage.

Judge C. C. Cummings.

The orator of the occasion, Judge Cummings, in a few preliminary remarks explained the map of the Battle of Gettysburg. There was considerable of the grim irony of fate in the Judge's remarks as he pointed out where the various generals and battalions were stationed on that eventful day. His empty sleeve hanging by his side, a relic of that dreadful conflict, gave silent evidence that the Judge knew what he was speaking about. It was a graceful act in the post to select one who had opposed them on that fateful day, and a wise act in the post to select a gentleman whose oration will add much that is splendid and valuable to the war literature of the country. Judge Cummings said:

Soldiers of the Grand Army of the Republic:

Let us glance once more at these lines when this charge was ordered. As senior sergeant of the Seventeenth regiment, Barksdale's brigade of Mississippians, I by order numbered the guns just before we started in, 417, in line. In three hours 275 went down in the storm. My little company B, had twenty-one, out of an original 100 more, not a man over twenty-five, but one married. By 7 o'clock seven were dead, seven wounded, two to death, and I and seven were alone of those left to tell how they died. They died, as in their boyish grandiloquence in camp and march they ever said they would thus.

"Whether the scaffold higher in battle's van.
The littlest place for man to die is where he dies a man."

When out in the wheat field you shattered our batteries with your shot and shell. These brave boys marched forward to support them and fell like grains before the scytheman's stroke. Volunteers must come and man again the wrecks of battered batteries. Then steadily they rose and went forward to certain death. Calm as a May morning were they in their "last full measure of devotion."

Prison Experience

I will now briefly sketch my 100 days of prison life at Chester on the Delaware, the oldest town in the Keystone state, and surely the most likely remembered of any north by those who were – 2000 of us

– so kindly cared for there by your good people. In the last days of June, 1863, I came down with the head of the column to the line of Mason and Dixon, at Middleburg – so called because the main street divided Maryland from Pennsylvania. I asked an old man to show me exactly where this line could be seen, about which a million had died and millions morn. He drew a mark in the center of the street and said it was just there. Then with a running bound I leaped over into Penn's woods and said: "Old man, I was something over there, but over here I am nothing but a freebooter." And so to carry out this resolve, that evening before nightfall, as we bivouacked on the Chambersburg pike I thought to sully for foraging. Soon a Dutch barn have in sight, and then a house, with its great brick oven nearby. The bar first, the brick oven next and the house last, in the order in which these thrifty people do build. I had formed a resolution most foul to charge the barnyard for fowl, but standing in the door, leaning against its facing, was a picture fair of beauty rare. It met my enchanted gaze. Imagine me repeating to her after the manner of Whittler, your Quaker poet, such a strain as this, as he has in his "Yankee Girl"

> "O, come where no winter thy footstep can wrong.
> But where flowers are blossoming all the yearlong,
> Where the shade of the palm tree is over my home,
> And the lemon and orange are white in their bloom."
>
> Then could you have seen her – that pride of your girls.
> Arise and cast back the dark wreath of her curls,
> With a scorn in her eye which the gazer could feel,
> And a glance like the sunshine that flashes on steel.
>
> Back, haughty Southron, your sky may be brighter than ours,
> And greener thy landscape and fairer the flowers,
> But dearer the blast round our mountains which raves,
> Then the sweet summer zephyr which breathes over slaves."

I went back to camp without the chickens. When the fourth day came, the fourth day of July too, I woke from a stupor of narcotics, at the field hospital near the Ridge, by the sound of the voice of the faithful black George, the friend of my childhood, boyhood, manhood, the best I ever had on earth, for he was soon to seal his devotion with his life to remain with me. All day long in the drizzling rain we trundled step by step the Chambersburg pike backward, and night found us only eight miles away, in Cashtown. The slow loss of blood all day caused me to leave the ambulance with him and seek shelter in a barn, to the left of the road, in this town. Next evening

George told me that your men were coming, and I insisted that he should go and enjoy his freedom, having served me to the end. He would not go, but wanted to gain the retreating column over the mountain side in the night, so that we might meet again. Then, before they got to us I had him take my sergeant's sword and put it in the side of the barn and break it off the hilt. They could get me but not that emblem of my honor. I can yet see the stubborn steel flashing back its anger in the fading light of day, bending stoutly to its death; but it snapped, and George went weeping in the gray twilight, scaled the mountains, and next day as your pushing advance column came upon him, he ran. They saw only the gray, not the black, and called halt. He would not, they fired and he died. Such is the ending of a domestic scene of southern slave life, and among the last. Could I ever hope to find such as devoted a man and brother?

For days in that barn there, I hovered over the confines of eternity. So close to death's door was I, that I thought I could hear it creek upon its hinges, and I would have gone, too, but for a good woman. Dutch she was. One day about noon I looked over into the meadow from my crevice in the barn toward the battleground, and saw the green pastures and the silver thread of its streamlet winding through, and the blue sky above and the birds singing in the apple trees around the barn. I strained to catch once again through the opening in the barn one more glance of that heaven that bends above, as Marie Antoinette did when she died. I longed to stay, for so young and too little of life had I seen, I was curious to see too great a silken cord we were unloosing. The freedom of all men everywhere. Yet I did not fear to go, for I had naught that my conscience did not approve, and was ready for the judgment. Then it was my blinding eyes caught the dim outlines of the tawny beard of a Pennsylvania Dutchman.

He had come with his frau to see the "repels" – there were thirty of us in that barn. Then my deafening ears heard a roar as of a wild beast: "Ah, you steal my horse, my hogs, my corn, you kills my boy over dare," pointing toward the battlefield, "and now you will die, and I am glad of it." With one supreme effort I raised myself on my elbow and flung back at him: "I will not die; I will live to fight you and your sort again." Then his tender hearted wife said "Hans, Hans, let the poor repel alone; vill you haf some apple butter?" I clutched it as I would prize of gold, for bacon and hard tack had shriveled up my soul and had dried up what little blood was left me. I gulped it down; it seemed to traverse the very marrow of my bones. For days and days she came and brought me apple butter and I was saved. God bless that Dutch woman and apple butter!

One day in August, the 18th your people came and took us up the pike to Gettysburg, Baltimore and up the Delaware river to Chester. The hospital was in the high school building at the edge of town, with ample wards attached. How very elegant was all. I seemed in a land of enchantment, so sudden was the change from the barn. Patent iron bedsteads, good grub and milk punch, snowy sheets and pillows, gas jets and chandeliers, with flags overall flags, on the wall, little flags and big flags, and young flags and the old flag, with

Stars and Stripes

Everywhere, too, were sketches from holy writ. One, especially, shall ever remain with me: "Whom He loveth He chasteneth." Surely, said I, he loves me, for now solely have I been chastened. One day in the reception room, where they had summoned me to come and met my parents from southland, I met a lady, a fair haired Saxon girl, with her father. They had a history He was a cavalier from the Palmetto state, married a lady of aristocratic art, and so Pierce Butler transferred Fannie Kemble, the renowned seductress, from the stage to his Carolina fields. This was an only child, their daughter, Fannie. Secession had divorced them. He had given up his country, She, entirely Union, wrote a book about the old flag, and the slave life she saw in the rice fields of Carolina. The daughter followed the fortunes of the father and was there with him to give his suffering people aid. It was whispered that the loyal surgeons went soft on fair Fannie. Be this as it may, certain it is she had full sway there, and did many, very glorious deeds of charity for suffering humanity, as well as done by many of your people that were not southern sympathizers. Pierce Butler, whose kinsman is now in our Senate, was gathered to his fathers. And Fannie, the fair, with the Saxon hair – what became of her? She fitted back with the little rice birds, that go north every season, to her rice fields. And when the war was at an end a titled Englishman found her wandering up and down them unmated and bore her away from us. They met by chance – the usual way. She is now over there, and ladies, I'll wager she, like Gladstone, are a good Home Ruler.

Off For Home. Then came the happy day when we were to steam for home, down the Delaware out into the broad Atlantic via Fortress Monroe, in the good ship City of New York, Captain Chisholm – kind soul, may he live forever and never die. Like the sailor boy, I saw the ocean, wild in all its glory, and sunrise on the green, green sea – not the blue sea, but green were its waters to me, I noted too, the white track of the ship with Mother Carey's chickens fitting on in the rear, following with an intelligence, humanlike, after gain from the droppings of the vessel. We passed the dark and frowning Fort

Monroe, at the entrance of the James, and next, with a shudder, the shadow of that old town of the James where landed that human cargo of twenty that brought us all these incidental woes. Next arose the shore at City Point, and hastening again to the land of my sires, I stood erect with hat in hand and exclaimed:

My Name is McGregor

and I stand upon my native hearth." Now, ladies and gentlemen and soldiers, you see I say soldiers apart from the gentle, for I doubt we be entitled to the term after all these wild works of woe and war, let us, too, have a drama with all these characters here therein. Let us call it the charlot of our new constellation, bearing therein our nation born again anew in its new birth of freedom, emblemed by Uncle Abe and Uncle Robert, as we see them here, side by side. We will have them holding the lines of the two grand forces, and we will have along Whitney and Lady Greene and Davis and Johnson and the antidote, Aunty Harriett, on whom we all done, and Uncle Tom and the faithful black George and we will traverse the world around in search of that lasting peace among ourselves and all nations, that Uncle Abe prayed on this field we might achieve. We will have, too the good master. But not the bad, but in his stead something worse. That doughty warrior who, like Job's horse of war, ever smelleth the battle from afar and paweth the valley because he was not there, and wherever fighting is to be done we will put him foremost. Then we may be assured that we will have that peace that Grant so long prayed to see, and he said he was joyed to know had come to us all before he stacked arms on the other shore. For, I venture, that if this valiant knight of the jaw-bone and wind bag but smelleth from afar a chance of war, as these honored dead here have known, he will shriek back and say that he too, has had enough. (1)

* * *

FORT WORTH DAILY GAZETTE
JUNE 20, 1891

C. C. CUMMINGS
17TH MISSISSIPPI INFANTRY

I was captured at Gettysburg, and being among the last relay of prisoners sent forward my experience at Chester, Penn., was in happy contrast to those at Fort Delaware and Point Lookout and other points. I offered to write this for the Century magazine as a bright exception to the general rule, and in justice to my Northern brothers, but the management had not the space, so I have shown that I am at least, impartial.

* * *

Fort Worth Daily Gazette
August 1, 1891

C. C. Cummings
17th Mississippi Infantry

Souvenir of Gettysburg.

Judge Cummings has received from Capt. W. B. Ford of the Fort Worth Fencibles, who is on his bridal tour, a catalogue and map of the cyclorama of the Battle of Gettysburg as exhibited at Philadelphia. It is dated July 2, 1891, on margin, as sent then by Capt. Ford. It happens to be the anniversary of the day Judge Cummings was wounded there in the peach orchard, just twenty-seven years from that date. It is a souvenir that many Gettysburg veterans in this city would appreciate, which, they can see by calling at Judge Cumming's office.

* * *

Caldwell (TX) News Chronicle
January 14, 1898

C. C. Cummings
17th Mississippi Infantry

The sudden death of Gen. Sul Ross was a shock to me and his other college mates here in Fort Worth - Henry Edrlegton, Oliver Kennedy and Joe Pankey. He and I were the same year and were in the same class for three years at Wesleyan University, Florence, Ala. Wes. Downs of Waco was in this class. The following Texas boys, then residents of Texas, were in the class below:

Oscar Downs, dead; Thos G. Davidson, dead; Haywood Brahan, dead; and Judge Joe Polley, living at Floresville, Texas, Brahan was for a long time identified in state politics under Goree and successor in the management of the penitentiary affairs. And I forgot to name Davis B. Gurley of Waco in the class above us all. Major Gurley and the Downs brothers were with Ross in the west, and followed him throughout all the viclasitudes of his romantic career as rough riders, and won fame eternal in asserting the right to attend to one's own affairs, and forcing others to do the same. Brahan and Polley were in Hood's brigade of Texans in Virginia, where I marched to victory with them through all these fierce fights till I left them at Gettysburg in the Peach Orchard, while they were climbing the rocky heights of Little Round Top and skulking the foe out of Devil's Den. Ask Jim Hasley, walking the streets here every day, he was there.

* * *

DUBLIN (TX) PROGRESS
AUGUST 8, 1913

C. C. CUMMINGS
17TH MISSISSIPPI INFANTRY

Gettysburg

(Mr. C.C. Cummings in the *Fort Worth Record*. Mr. Cummings has addressed the Confederate Veterans of Dublin.)

The night of June 20th we bivouacked near Cashtown, about six miles from Gettysburg. All that day on the march I had presentiment of evil to befall me in the coming battle. Just where the clash of arms was to be we knew not, but every private was as much aware that it was going to be the decisive battle of the great war as any general. We crossed over into the enemy's country seeking to end it at a blow and Gettysburg went down in history as the field, the fatal Flodden field. All the day of the 30th I was obsessed with this presentiment of evil and I confided my forecast to Lieutenant Colonel Fizer, saying I preferred a wound in the arm that I might escape being shot again after being down, which frequently happened when wounded in the locomotive powers. Fizer was a dressy comrade and preferred a wound in the leg, as he did not wish to mar his shape. We both got our preference on the second day in the peach orchard where we went up against Sickles' corps in that valley of death. That night of the second, there was gathered in a tent in a field hospital on the grounds all the field and staff of our regiment, the seventeenth Mississippi, except Dick Jones, the adjutant killed on the field. Colonel Holder was there with a wound in the groin; Fizer with a wound in the leg, as he preferred, and the writer, sergeant major, in the hand, as I preferred; Major Pulliam also in the hand, and Brown Jones, the Colonel's orderly, about fifteen in the body. We bivouacked at Willoughby Run on July 1st, where was opened the first day's battle and where a monument has been erected to the memory of the Federal General Reynolds, killed there that day. I claim to be the original reunionist, of the Gray and Blue. For when I was shot in the hand and while the blood was filling my buck gauntlet (I put on my best to die like a soldier,) two New York Zouaves running my way to avoid capture by our gray boys, who had broken through a barn on my left, and who would have served them roughly for the resistance they made. I beckoned them to come to me and I flung my arm around the neck of each, making a wounded squad, which our long experience told us would save us from being fired upon, and so I came off the field with my two prisoners. Wouldn't it be passing strange if I should meet one or both after fifty years at Gettysburg? Sickles will be there and they were Sickles men. Sickles and I exchanged compliments that day. I gave him my hand and pulled his leg. All the third day of July, 1863, we lay prone of our wounds in the hospital tent on the field of Gettysburg and heard the thunder of this legions or guns – big guns – from both sides, such guns as were never heard on the American continent. It was what has gone into history as "Pickett's charge" and while there were others, who crossed the deadline with Pickett to the high water mark, yet old Virginia our dear old mother, bore on her sacred bosom the bosom of war for us all. So let her son Pickett go down in history as the hero of us all.

* * *

DENTON RECORD-CHRONICLE
OCTOBER 13, 1913

C. C. CUMMINGS
17TH MISSISSIPPI INFANTRY

Judge Cummings Spoke At Church "Gettysburg," Sixteenth Decisive Battle of World's History." Interesting Subject Of Gettysburg Veteran's Address Here Sunday.

The address of Judge C. C. Cummings of Fort Worth, Confederate historian for the U. C. V.'s was delivered to an audience which fairly well crowded the First Christian Church on Hickory Street, Sunday afternoon. Subject of the talk "Gettysburg, the Sixteenth Decisive Battle of the World," made it one doubly interesting to the veterans, recalling to many of them who had been with Lee in the struggle between the "two ridges" of the little Dutch town all the fierce hope of victory with which the first three days fifty years ago had begun, and the bitter certainty of defeat which followed the second day's battle. The church well filled after the Confederate veterans marched in, after the flag, to places reserved for them and the young men of the Reagan and Lee societies of the North Texas State Normal followed.

After announcing his subject, and a short explanation of it, Judge Cummings spied Dr. J. R. Edwards in the audience and asked him to come up to the platform, laughingly desiring him to act as "witness" to the story of the Battle of Gettysburg, where he and the speaker were comrades. It was a rule made in Lee camp, he said, never to tell a theory unless you had provided yourself with a witness, after which statement he extracted a promise of unrestrained endorsement from Dr. Edwards.

Tells of 1913 Re-Union

A short delineation of the pleasures of the recent Gettysburg celebration, which Pennsylvania as hostess to the boys in blue and gray, was a great pleasure to many of the local veterans who were unable to attend the "peace" reunion. Their greeting at Gettysburg, he declared, was as warm as it had been on a memorable day fifty years ago, though of a deferent sort. Beginning with his story of the great battle in which 72,000 Confederates and 83,000 Federals were engaged, the speaker introduced a map of the city of Gettysburg, pointing out the places concerned as he told them. He followed the movements of Lee's Army of Northern Virginia from Chancellorsville, where Jackson, "the right hand of Lee," was mortally wounded and died, to the assembling of the army at Culpepper's Heights on the eighth of June, with Gen. Hooker just across the creek.. On June 25 Lee crossed the Potomac, Hooker having crossed at Edward's Ferry previously. Lee, whose plan was to move up into the enemy's country, came on to Chambersburg, and sent Gen. Gordon on ahead to threaten Washington. Lee halted, and he learned at Chambersburg that Hooker was in the neighborhood on the line above. Stewart was not successful in crossing the Potomac to get news of the movement of Hooker's army. The speaker digressed a little to compare the Battle of Gettysburg with the fifteen other decisive battles of the world, saying that while at Waterloo, where Napoleon met his defeat, there were about the same numbers engaged and last in the battles that the great emperor was effectively crushed with the defeat on the battleground of Waterloo, while after the Battle of Gettysburg, Lee calmly turned and stood ready to meet Meade, should an attack be made.

Compliments Gen. Meade

The speaker complimented the command of Gen. Meade, who was in the charge of the army three days, following the removal of Gen. Hooker. He told of Lee's orders to his troops to assemble at Cashtown, meaning to form the base of his operations at the foot of the Blue Ridge mountains, Rhode's division developed fight, and Lee ordered the Cashtown troops forward to Gettysburg and Gordon's troops, far in advance, back to the small city.

Then followed an intensely interesting description of the three days' battle itself, with a description of "the most beautiful battleground in the world," then the speaker pointing out each place and movement on the map of Gettysburg.

The speaker made himself very clear on the point of his right to the title of "Confederate" soldier, declaring that the surviving men in the old ranks of gray were Confederates still, and only "ex-soldiers. That they had fought for what they "believed" to be right was his emphatic statement, without bitterness. That the great rebellion was as necessary to the future cementing of the Union as any other agent, he believed, and blood, he said, was the only cement that would answer.

Giving the definition of history from a modern encyclopedia, he amended it by an edition. "History is the science of the progressive development of the human race," "under the guidance," he amended, "of a Supreme Power."

The speaker spoke for a short time of his personal experience in the war, the loss of his right hand, which came about when he "exchanged courtesies" with a Federal soldier, giving him a right hand for left leg. His capture at Cashtown and three or four months imprisonment, and other incidents of his personal experience made interesting telling.

* * *

FORT WORTH DAILY GAZETTE
JUNE 16, 1894

C. C. CUMMINGS
17TH MISSISSIPPI INFANTRY

A Faithful Slave Who Followed His Master Unto Death

Editor Gazette,

Col. J. W. Friend's article, "Old Sol," about the good old darkey of his mess "endurin ob de wah" is a timely remembrance of the faithful man in black, who followed our fortunes in the "storm cradled nation that fell" as little children, and whose kind, simple, humble souls will never be equaled by another such race in another such patriarchal age. I too would add another tribute to the memory of another such who was with me from the first battle of Manassas till the Battle of Gettysburg, where he lost his life for his friends. The Mater of the human heart says greater love hath no man than this. My father - Kentucky born - brought him from Kentucky and we called him Kentucky George to distinguish him from another on the place of that name. Early in April, 1861, I enlisted when the tocsin sounded for home defense and George was my constant companion till his death. We were about the same age and never in my life have I met a character, white or black, whose gentle humility of spirit came so near the standard as set up by Christ as this poor, kind, sable-skinned son of

Africa. He was not afraid to fight either. At the battle of Leesburg in October, 1861, noted for being the death of General Baker, Lincoln's friend and for whose Lincoln mourned so sincerely, when my regiment (Seventeenth Mississippi) rushed in about sun down and decided the fight, we ran over two howitzers which George assisted to carry in triumph from the field. But he was more valuable to us as a cook and forager than a fighter, so we were careful to keep him in rear. When wounded at Gettysburg I had orders from our surgeon to remain and be captured or a move would disrupt the artery just taken up. But when the army began to move back on that morning of the 4th of July, 1863, I called to George to get me an ambulance which he did and all day we moved back on the pike in that slow drizzling rain and when night came we had only got eight miles at Cashtown. Sure enough, the artery broke and I had to stop or bleed to death. That night we slept under the same blanket - George and I - on the porch of a Pennsylvania Dutchman and in the morning repaired to a barn in an apple orchard opposite. About sundown George told me the Yankees were coming and had begun to separate all the black folks from their masters. I told him his freedom he had and earned and was his, but he did not want it that way, and would not have it so. He determined to go back over the mountains that night and join the regiment and wait for me to be exchanged.

I told him then to go as he desired, but to first break my sword so the Yankees could not get it.

I was sergeant major of the regiment. Its gleam and glint in the dying sunlight of that woeful day as it broke in twain to the crack of that barn is yet as vivid as yesterday. I was in prison and never joined my regiment after being exchanged and never knew the fate of poor faithful George till about a year after the end of the war. He had made it safely over the mountains and was nearly of our forces when he was stopped by a Northern lady to enquire about her son whom George happened to know in our army. He had just told her that her son was safe and a prisoner when the advance of the Yankees came up and seeing the gray jacket began firing on, it and George was killed. The lady saw him buried and brought the story over the lines.

C. C. Cummings

* * *

JACKSBORO (TX) NEWS
JULY 8, 1914

C. C. CUMMINGS
17TH MISSISSIPPI INFANTRY

Judge Cummings Address

A small audience at the courthouse on July 4, under the auspices of the local Confederate camp here, J. C. Jones, commander, who followed by an earnest appeal for better and more efficient camp organization.

Judge Cummings main topic was the Fourth of July in History as appropriate to the day. He went back to colonial times when Geo. Washington, a youth of 21, led a party of Virginians against a force of French who were entrenched at what is now Pittsburg and on the 4th of July 1753 not far from where a battle occurred, the battle of great moment in which Washington was compelled to surrender to a superior force. Then twenty three years later after this the 4th of July 1776 came our declaration of independence. Then fifty years later this July 4th, 1826 Jefferson and Adams died, both of whom had signed the immortal document and Adams became second and Jefferson third president of the United States.

The speaker then referred to the remarkable coincidences of what occurred on the 4th of July 1863. After three days battles 1st, 2nd and 3rd of July at Gettysburg, the next day the 4th of July found Lee slowly retiring from Gettysburg, on the same day Pemberton surrendered at Vicksburg. The Confederate General Holms with 10,000 of his men was repulsed in an assault on Helena and on the 4th of July at the same time Morgan was captured in his raid in Ohio and in the same day Port Hudson was preparing to surrender and did a day or so after and there was the speaker said the bottom dropped out of the Confederacy and began slowly to settle its doom.

Then he referred to the fact that just fifty years after the Gettysburg battle 50,000 veterans of Blue and Gray met again on this bloody field as friends and four days fraternized as guest of Pennsylvania and Uncle Sam who joined in the immense expense of feeding and tending to this great host. Judge Cummings was among the visitors and reviewed the ground where his command charged down in the Peach Orchard, July 2, 1863 and where he lost his right hand.

The audience, though not large, seemed greatly entertained among whom were many young men who have come on the stage of action in a later generation. He alluded to the fact of the crus- fixation of the South on the Southern battle flag and aid the South, though dead and buried, had risen again with healing in its wings with a Virginian and for president, a descendent of Virginia vice president, a grandson of a Virginian as Secretary of State, a native Texan as Post Master General, a native of South Carolina as Secretary of Agriculture, a native of North Carolina, secretary of the mightiest navy of the world, a Tennessean as attorney general, a Confederate soldier at the head of the Supreme Court; the most august tribunal of the earth; a native Georgian secretary of treasury; the president lately declaring for states' rights in all cases where the issued can be managed better than the government in mass. He alluded to his attendance at the most recent reunion at Jacksonville, Florida and that De Soto's conquest of Florida in 1542 resulted in embracing the seven gulf states beginning with Florida and ending with Texas was for more than a hundred years after all called Florida. That on February 4, 1861, at Montgomery, Alabama, in the center of these seven Gulf States raised the Stars and Bars which seven states embraced these seven states and that now the canal was opened these seven states would control the commerce of the world, just as the seven stars lead the heavenly bodies in their course and that Texas was to be the lead of all the 48 stars in the galaxy of our charming circle.

* * *

WEEKLY CLARION-LEDGER
OCTOBER 22, 1903

PRIVATE FENTON L. FULGHAM
18TH MISSISSIPPI INFANTRY

[The writer] wound up his military career at Gettysburg in that memorable charge of Longstreet's corps, on the 2d of July, 1863. There is a small number of survivors of company E 18th Miss, who were participants and suffered the pangs of defeat and shard the glory of victory in that eventful period of our country's history. Among those were Capt. Billy Lewis, of Clinton, W. J. Green and Elisha Roberts of Copiah, John Brent of Raymond and Lewis Terrel "Old Red" who now lives between Jackson and Clinton. They are all about of Co. E. that I can recall, there may be others.

* * *

MACON (GA) BEACON
MAY 8, 1914

CAPTAIN T. J. STOKES
11TH MISSISSIPPI INFANTRY

As soon as we reached Fredericksburg we started on our march to Pennsylvania. On the morning of July 1 commenced the firing which ended July 3 at Gettysburg. Our regiment was left July 1 to guard the wagon train at Cashtown. On the night of July 2 we reached the main line at Gettysburg. On the evening of July 3 we were in the main attack on the Federal army, where I was slightly wounded and captured. On the Sunday after the battle I was carried with a number of prisoners to Baltimore where I, with other officers, was placed in prison at Fort McHenry. Remaining there a few days, we were taken to Fort Delaware. Remaining here about a week, we were carried to Johnson's island, near Sandusky, Ohio, which place we reached the latter part of July 1863. I was in prison until March 7, 1865, when I was exchanged and sent to Richmond.

* * *

YAZOO HERALD
MAY 15, 1908.

FIRST SERGEANT WARREN D. REID
11TH MISSISSIPPI INFANTRY

I am indebted to two veterans "who wore the gray" for valuable papers on the great battle of Gettysburg – one of them Mr. Warren D. Reid of Houston, Miss. He belonged to Company H, 11th Mississippi Regiment. The other one is Mr. McNamara of Wilkerson county, whose home is now at Beauvoir. He belonged to the 21st Mississippi Regiment. These two soldiers could not have written of a more desperate battle, or any other where Mississippi troops behaved with more superb courage than here at Gettysburg, called by some writer, "The Waterloo of the Confederacy." Gen. John B. Gordon said Gettysburg and Appomattox fix the boundaries of the Confederacy's decline and death. At Gettysburg its sun reached its zenith and passed its meridian; at Appomattox it went down forever. The mighty hosts had met in the fierce death grapple. The supreme struggle at Gettysburg was at hand. Thee it was the 21st Mississippi Regiment bore the stars and bars to the farthest point reached on that bloody battlefield. We learn from one who knew, that "the Seventeenth United States Regulars lost 62 per cent in killed and wounded in the Peach Orchard, and upon the ridge beyond. The state of Massachusetts has erected a magnificent monument to mark the spot where Bigelow's Battery was annihilated. There is no stone there to tell this story, but the fact remains that the regiment that swept through that Peach Orchard on Gettysburg's crimson field and up the height beyond, mowing down the 17th U. S. Infantry and destroying that battery, was led by a Mississippian, a son of Claiborne county."

Members of the 21st Mississippi Infantry postwar

(Old Statehouse Museum, Vicksburg, Mississippi)

Chapter Six

Mississippi Casualties at Gettysburg

BIRMINGHAM NEWS
DECEMBER 7, 1924

PRIVATE L. BRITTON
42ND MISSISSIPPI INFANTRY

The journal of Capt. Wollard relating the mortality of this company in the famous charge of Gen. Pickett at the battle of Gettysburg is a sad one. Only two men escaped death, wounds or capture. Capt. Wollard was captured in this charge after being wounded. The company went into the charge 100 strong. Among those who died, there was L. Britton, a boy only 14 years old.

* * *

GREENWOOD ENTERPRISE
MAY 23, 1902

PRIVATE FRANK A. HOWELL
11TH MISSISSIPPI INFANTRY

It gives us satisfaction to publish the following letter from a gallant veteran in correction of one of our statements in last week's issue relative to the heroes who fell on Cemetery Heights. It is our desire that all participants who remembered the facts with accuracy would write us and set right any possible errors; hence we publish Adjutant Howell's letter gladly:

"U. C. V. Holmes County Camp No. 398 Bowling Green, Miss., May 9.

"Dear Miss Power – in your comments on what the *Dallas Times-Herald* says about the troops that went highest up Cemetery Heights at Gettysburg, July 3, 1863, you justly say that Davis Mississippi brigade went, as their dead bodies showed, as far as the farthest: but

*Private Frank A. Howell
11th Mississippi Infantry*

(Confederate Veteran Magazine)

---

you are in error as to one regiment that composed of the Second, Fourth, Eleventh and Forty-second Mississippi Regiments. At its first organization it had the three first-named regiments and the North Carolina regiment, but when the Forty-second Mississippi was formed it was put in Davis' brigade and the Carolinians went into Pender's North Carolina brigade. Both were in the charge that afternoon. I carry until now a short leg, caused by a minie ball received in that charge. Statistics also show that the Eleventh Mississippi Regiment suffered a greater proportional loss on that day than any Mississippi Regiment and I believe any other regiment, except possibly a North Carolina regiment. Of course, a private's scope of observation was limited, but I had a brother captured that day who went as far as any troops, and he had a comrade that was taken prisoner at the same time and place who is now living in Starkville, Miss. (James T. Jones), who can give living testimony to their advanced position when captured. I only write this to correct your statement as to what troops were at that time in Joe Davis' Mississippi brigade. I was until that day a member of Company F, Eleventh Mississippi Regiment. We took forty-two men into action and only five answered roll call next morning; eleven were killed or mortally wounded. Yours truly.

"F. A. Howell."

* * *

QUOTED IN: THE CONFEDERATE WAR
CLARION-LEDGER

Private Frank. H. Foote
48th Mississippi Infantry

We often see statements of the several State losses of the Confederate troops of the battle of Gettysburg fought July, 1863, and much controversy exists right now as to which State suffered the heaviest loss, particularly between Virginia and North Carolina. No one seems to have cast his eye over the list of States which suffered the most and make an average and then see where the heaviest proportion lies. In looking over Surgeon-General Guild's report we find that contrary heretofore expressed estimates that the State of Mississippi suffered the heaviest loss of any in proportion to the number of regiments engaged. I take a list of six States which suffered the most and show the estimate which was a surprise to me and doubtless will be to others:

State	reg.	loss	average
North Carolina	33	3.947	120
Georgia	36	2,396	67
Virginia	40	2,283	47
Mississippi	11	1,350	123
Alabama	17	1,342	79
South Carolina	11	1,160	106

North Carolina, Georgia and Virginia claim to have done the fighting but the records show that Mississippi with only 11 regiments, suffered a heavier loss than North Carolina with 33, Georgia with 36, and Virginia with 40. So on through the whole war, when the fighting was heaviest and death the greatest, there you could and Mississippians. They have no historian to record their valor, nor do they claim any honor only what they justly and nobly earned. The figures speak for themselves.

Frank H. Foote.

Late 48th Mississippi.

* * *

Clarion Ledger
January 17, 1889

Doctor B. F. Ward
11th Mississippi Infantry

The Gettysburg Fight

Number of Confederate Soldiers Wounded There. Dr. B. F. Ward, Who was on the Field, Estimate the Number Between 7,000 and 8.000 – Strength of Lee's and Meade's Armies.

Editor of the *Clarion Ledger*:

Gentlemen – I beg leave to offer you for publication, the following statement of Lieutenant Bond, clipped from the *Times-Democrat*, also a report of the Confederate wounded taken from the *New York Tribune*, while on the battle-field of Gettysburg, where I remained for three weeks, under an order from Gen. Heth, in charge of the wounded of his division.

These statements are presented with no purpose of entering into or encouraging controversy as to the comparative merits of the different commands engaged in that great struggle, but simply to furnish such of your readers as may be interested in these matters, a few historical figures.

The *Tribune's* report, as far as it goes, is literally correct because it is taken from the Federal authorities, we were required to make every day, duplicate requisitions, accompanied by a report of the wounded on hand, reducing the list, each day, by the number that had died or been sent forward to Federal hospital.

The number of wounded accredited to Heth's Division, by the Tribune reporter, is exactly the number on which I based my first requisition for supplies the day after Gen. Lee's army retired. It must be remembered, however, that only a limited number of our wounded fell into the hands of the enemy. Nearly all who were able to work, or bear transportation in wagons and ambulances, went back with the army to Virginia, leaving on the field only those who were too severely wounded to be moved. I think it would be safe to estimate that more than half the wounded of Heth's Division went to the rear with the retreating army. I have not preserved a list of the casualties sustained by the 11th Miss., in that battle, but there are doubtless surviving members of that famous regiment, now in the State, who could approximate its very heavy loss.

According to my recollection it came out of the fight under the command of Lieutenant Stephen Moore, of Aberdeen; all the field officers and captain having been either killed or disabled. There may be a few typographical errors as to names of division commanders and surgeons, but the figures are correct.

Dr. Benjamin F. Ward of the 11th Mississippi Infantry

(Beth Austin)

New York Tribune, July - 1863.

The Rebel Hospitals

The rebels established their hospitals on the Southwest and Northwest of the town, in the Pennsylvania College, the Seminary and in all the private houses on the roads by which they marched. There are great numbers at or near Cashtown. Many of these we would not have discovered at once had they not been so destitute that they were obliged to visit us for supplies as our army countermarched to head the enemy of instead of chasing him in the rear.

Rebels wounded in Hospitals near Gettysburg, to whom rations have been issued.

Johnson's Div, under	Surg. Whitehead,	311
Rhode's " " " "	Hayes,	800
Hood's " " " "	Means,	515
Early's " " " "	Gott,	259
Anderson's " " "	Miles,	111
Pickett's " " " "	Reeves,	279
Parson's " " " "	Ass't Surg. Wilson,	171
Pender's " " " "	Surg. McAdams,	700
McLaws " " " "	Patterson,	700
Johnson's " " " "	Sayers,	135
Under charge of Surgeon Frazier's,		700

Total 5374

Of these there have been sent forward to date about 2,000

Leaving on hand in these hospitals 3,374

The above are of course in addition to the two or three thousand rebels wounded in our hospitals.

Thus, according to these figures, which are, perhaps, the next accurate that have ever been published, Gen'l Lee left on that fatal field between 7,000 and 9,000 badly wounded.

His whole force was 62,000 men. When he recrossed the Potomac his army had been diminished by 19,000 killed, wounded and missing. Gen'l Meade's force was 112,000; his losses, according to his own report, 24,000. He stated, in effect, before the committee on the conduct of the war, that if the position of the two armies had been reversed and his had been the assaulting instead of the fortified and defending force, that success would likely have been with Gen'l Lee and defeat with him. God only knows who was right, but when the coming generations shall have been far enough removed from the scenes of this fearful conflict to enable them to calmly study the philosophy of history instead of its bare and bloody skeleton, the North and the South will learn to mutually honor and appreciate the respective valor and patriotism which makes them a brotherhood of heroes and a sisterhood of equal and sovereign States. B. F. Ward.

W. B. Bond, late First Lieutenant and A. D. C., Daniel's Brigade, Rhodes' Division, Confederate Army, writes to the *New York Sun* from Scotland Neck, N. C., as follows:

Gen. Longstreet, in his speech at Gettysburg on the 2d instant, referring to his assault on the third day, praises Pickett's and Trimble's troops, but carefully ignores Heth's. Why is this? All soldiers know that the number of killed is, the one and only test of pluck and endurance. Besides five North Carolina and three Mississippi Regiments there were troops from Tennessee, Alabama and Virginia in Heth's Division, and all of them behaved gallantly except the left brigade, which was Brockenbrough's Virginia, and its loss in killed in this battle was twenty-five or five to the regiment. Pickett's fifteen (Virginia) regiments had 224, or average of nearly fifteen to the regiment. I have never seen the casualties of the 11th Mississippi, but the number of killed in the Second and Forty-second Mississippi added to that of the five North Carolina regiments was 338 and an average to the regiments of something over 48. The fire of Cemetery Hill was concentrated upon Heth's Division, and at the close of the charge its organization was to a great extent broken up; but, with the exception of the left brigade, this disorganization was caused by deaths and wounds. Pickett's Division, when 1,500 of them surrendered in an open field was better organized, but this fact was owing entirely to their comparatively small loss in killed and wounded. The figures used are official. I was at Gettysburg and was severely wounded.

* * *

DEMOCRATIC-HERALD (CHARLESTON, MS)
APRIL 4, 1901

ALLEN J. HOOKER

Barksdale's, Posey's and Davis' Brigades at Gettysburg.

The following correspondence and data make an important contribution to State history, and will be of special interest to the survivors of the commands named:

Gettysburg Pa., March, 9, 1901.

My Dear Col. Power – As the Confederate member of the Gettysburg Battlefield and Park Commission it falls to my lot to prepare the inscriptions to be placed on the historical tablets which the government is erecting to the Confederate troops on this field. Can you procure and send to me the number of men in that brigade at Gettysburg? I shall be greatly obliged to have this soon, for we wish to erect the tablet to that brigade, among others, this spring. It contained four regiments – the 13th, 17th, 18th and 21st.

I should like, also, to have the number at Gettysburg, in Posey's brigade, composed of the 12th, 16th, 19th and 48th regiments. And also the number in the 2d, 11th and 42d regiments of Davis' brigades (which also had the 55th North Carolina in it, whose number I have.)

No troops were better or braver than the Mississippians. I sued to fight with the 11th regiments in Bee's and Law's brigades. Yours truly,

Wm. M. Robbins, Maj. 4th Ala.

None of the official reports of Gettysburg give the total strength of the Mississippi brigades as they entered the fight. But very fortunately we have a copy of the *Mobile Evening News* of Sept. 11, 1863, that contains the desired information in the following very interesting communication.

Enterprise, Miss., Sept. 8, 1863.

Mr. Editor: The following circumstances attending the fall of Brig.-Gen. William Barksdale at the battle of Gettysburg, related by his adjutant general, Captain John A. Barksdale, will be of interest to his friends:

"Gen. Barksdale was struck twice – first in the shoulder and afterward in the chest. When last wounded he said to Boyd (a courier), 'I am killed.' Tell my wife and children that I died fighting at my post.' Just at this time Boyd was wounded in the leg, which made it impossible for him to render Gen. Barksdale any assistance. He, therefore, left him, the enemy not being more than twenty yards distant. The cause of my not being with him at the time he was shot is, that about fifty yards from where he was struck my horse was killed under me. I was with him at the time and remarked to him that I was on foot. He told me to go back to a battery, about 250 yards, and order it to the front. This I did, and returned as rapidly as possible. When within forty yards of the brigade I saw Gen. Barksdale's horse galloping over the field without a rider, and the men falling back, while the enemy was advancing. To go to him was impossible."

The last words of this ardent patriot, gallant man and dying hero that ever fell upon the ears of his own countrymen should serve and incite him to his own high standard of duty * * * The ordeal through which his brigade passed in that fight may be judged when it is told that of 1420 bayonets carried into it, 730 were lost in killed, wounded and missing.

W. R. B.

"W. R. B. are the initials of Wm. R. Barksdale, and he doubtless obtained the figures quoted in the letter from which the foregoing extract was copied.

Having referred Maj. Robbins' letter to Mr. Allen J. Hooker, he says:

The official reports do not give the numbers in Mississippi brigades. It is possible that none of the survivors may be able to furnish this. I copy the casualties at Gettysburg:

Barksdale's Brigade.

Killed, staff, 1.

13th Mississippi: Killed 28, wounded 137 – total, 165.

17th Mississippi: Killed 40, wounded 160 – total, 200.

18th Mississippi: Killed 18, wounded 82 – total, 100.

21st Mississippi: Killed 16, wounded 87 – total, 103.

Grand totals: Killed 105, wounded 550, captured and missing 92; total 747. Engaged in the battle of July 2d.

Posey's Brigade.

12th Mississippi: Killed 0, wounded 7 – total 7.

16th Mississippi: Killed 2, wounded 17 total 19.

19th Mississippi: Killed 4, wounded 23 – total 27.

48th Mississippi: Killed 6, wounded 24 – total 30.

Grand totals: Killed 12, wounded 71 total 83.

Davis' Brigade.

2d Mississippi: Killed 40, wounded 183 – total, 232.

11th Mississippi:	Killed 32,	wounded 170 – total, 202.
42d Mississippi:	Killed 60,	wounded 205 – total 265.
55th North Carolina:	Killed 39,	wounded 159 – total 198.
Grand totals:	Killed 180,	wounded 717 total 897.

This brigade was engaged first and third days, July 1 and 3.

The brigade casualties at Gettysburg give Mississippi second and fourth places:

1. Pettigrew's North Carolinians.
2. Davis' Mississippians.
3. Daniels' North Carolinians.
4. Barksdale's Mississippians.

Dr. J. R. Prince, Captain 11th Mississippi, in a letter to this writer, states that Lieut. Hester, who was captured at the stone wall, informed him that during his imprisonment a reputable Federal officer told him (Hester) that the dead of Mississippi were found higher up the heights than those of any other command.

Dr. Prince wrote that he himself gave the order to retreat on July 3, 1863, ever commissioned officer having been shot down, and that the brigade retreated in passably good order, about 300 in number, back to its original position on Missionary Ridge, occupied by it in the morning. So that if your correspondent will deduct the number of the 55th North Carolina (which he has) from the total, he can approximate the strength of the other three regiments just prior to the so-called "Pickett's charge."

* * *

EXCERPT FROM: MAUD BROWN, THE UNIVERSITY GREYS, 46

DOCTOR BENJAMIN F. WARD
11TH MISSISSIPPI INFANTRY

Gettysburg --- A Bit of History

(By Dr. B. F. Ward)

A great deal has been written about Gettysburg by the people who were not there. I was there, but, like any other man know very little about it so far as material facts came under individual observations. I was not in that battle and I am glad of it, because I am thankful of the privilege of having lived 50 years after it was fought. I enlisted in the Confederate service in May, 1861, as a private in Company K, Eleventh Mississippi Regiment commanded by that prince of soldiers, Col. P. F. Liddell.

At the battle of Sharpsburg in Maryland, Col. Liddell was killed, as was also Lieut. Col. O. Butler and Maj Evans, leaving the regiment without a field officer. The regimental loss in this fight was over 150 killed and wounded. At the end of the wet and weary march from Yorktown to Richmond, while standing in the ranks footsore and muddy, the adjutant by the regiment, Capt. Joe Evans handed me a commission. As full surgeon in the Confederate States army and ordering me to report to Gen. Joseph E. Johnston for duty.

Now as to Gettysburg, it is history that on the first day Lee encountered only the First and Eleventh corps of Meade's army.

Gen. Reynolds was killed and his corps routed and driven through the town of Gettysburg. The Confederates were having it all their own way. There were still two hours of daylight in which to pursue the panic stricken foe and occupy Cemetery Hill. Why did they stop? It was the fortune of war, because as a writer on the Federal said, "Jackson was dead."

I remained on the battlefield three weeks after Gen. Lee had retired and the Federals told me that they came into those heights at midnight, as they expressed it, with their "tongues out." They could not understand why we had halted. Neither can anyone else, except that "Jackson was dead." We lost the second days battle - after A. P. Hill's crushed Sickles in the bloody wheat field and shot his leg off - because of Longstreet's tardiness in failing to seize Little Round Top while it was still unoccupied by the Federals. Gen. Warren, Meade's chief engineer, said that if Longstreet had been even 30 minutes earlier he could have taken Little Round Top without opposition, which would have enabled him to enfilade Meade's line and would have saved us the day.

Again it was the fortune of war against us, because "Jackson was dead."

If Jackson had lived Lee would have won Gettysburg. If Meade's army had been vanquished and shattered, Lee would have marshalled his victorious legions at the foot of Washington monument and the Southern Confederacy would have been the model republic of the world. But it was the fortune of war that this should be on nation after "Jackson was dead." In a recent magazine article Gen. Nelson A. Miles, who put the irons on Mr. Davis, practically admits that Lee failed at Gettysburg because "Jackson was dead," and says if Lee would surely have taken Washington, and that it would have then been impossible to have organized another army in the North, because the anti-war spirit was just at that time reached its highest time.

Altogether Gettysburg was not up to be the standard of Lee's splendid skill and execution in all of his previous great battles.

As soon as he was recrossed the Potomac he tendered his resignation to President Davis and stated as one of his reasons that as he had not satisfied himself, it was fair to assume that he had not satisfied others, and that having "reached an age where he had to use the eyes of others," he felt that a younger man might accomplish more. When he said "it was all my fault," he was simply giving expression to the generous impulses of his great soul. It was not his fault. Napoleon said he could not make men, but had he to use such as he found. So it was with Lee. Lee was the greatest man of all the military chieftains in history, but not the greatest general. Caesar and Napoleon were his superiors and Washington was his equal. Lee. Jackson, Sidney Johnston and Forrest constitute the granite pillars upon which will test the temple of modern military science and achievement, so far as evolved by the Confederate war.

Next to those in force and efficiency was John B. Hood, who was a martyr to the fortunes of war.

The Federal, as well as the Confederate army, had a great many very able subordinate officers, but the North furnished no man who rose about the level of a capable commander of a large army with unlimited and inexhaustible resources – against a small army with limited and exhaustible resources. – against a small army with limited and exhaustible resources. This will be the verdict of history when the generation have receded far enough from the scene of conflict to escape all bias and all color of sentiment.

The generation born since the war has been thoroughly imbued with the conviction that "Pickett's Charge" held the front of the stage and was the crowning act of glory in the tragic drama at Gettysburg. "Pickett's Charge" is a historical myth insofar as it conveys the impression that Pickett charged further, more desperately, suffered more, or accomplished more than most of the divisions who participated in that desperate and deadly assault. Picket made one simple charge on the evening of the last day. This was his only effort. Other divisions had been charging for three days, though heat and dust and smoke and flame and blood.

His division behaved, as well as any of the others, certainly no better. In the heraldry of battles the list of casualties has been generally accepted as the highest test of courage and endurance. Fortunately for the truth of history, we are able to measure "Pickett's Charge" of all the others by this standard. When Lee was preparing to retire, Gen

Heth sent me a written order to remain there as surgeon in chief of all his wounded left on the field 600 in number, lying on the ground without food or shelter.

Let it be remembered that these were all badly wounded, many of them fatally, as those who were slightly wounded and could walk or bear transportation in wagons and ambulances went to the rear with the army. Of course medical officers and nurses were left with the wounded of each brigade in the division.

The brigade surgeons were required to report every evening to the chief surgeon the number of wounded on hand, estimating each day the number died, or sent off to Federal hospitals during the day. These reports were compiled every night by the chief surgeon and forwarded, in duplicate, next morning to Federal headquarters, as a basis on which rations and medical supplies would be issued to us.

This I did every day for three weeks, and this order, under which I was acting applied to every other division. It so happened that I was the only surgeon in Lee's army – s far as I knew – who took the trouble to secure and preserve these reports from all the Confederate surgeons exactly as they were made to the Federal authorities. I carried them in my pocket during the five months of my imprisonment and have kept them to this day. They speak for themselves.

Rhodes division	800
Pender's division	700
McLaws division	700
Hood's division	515
Heth's division	693
Johnson's division	311
Pickett's division	279
Early's division	259
Parson's division	171
Johnson's division	135
Anderson's division	111
Pennsylvania College	700
Total	5, 374

In addition to this the Federals claimed that they picked up two or three thousand of our wounded and carried them to their own field hospitals. Thus, according to these figures, which are perhaps the most accurate that have ever been published between 7,000 and 8,000 were wounded. Lee's whole force was 62,000 men. When he recrossed the Potomac his army had been diminished by 19,000 killed, wounded and missing.

Gen. Meade's force was 120,000, his losses, according to his own report, were 24,000. He stated, in effect, before the committee had been reserved and his had been the assaulting instead of the fortified and defending force, that success would likely been with Gen. Lee and defeat with him.

W. B. Bond, late first lieutenant and A. D. C. Daniel's brigade, Rhodes division, Confederate army, writing to the *New York Sun* in Scotland Neck, N. C., said:

"Gen. Longstreet, in his speech at Gettysburg - referring to his assault on the third day - praises Pickett's and Trimble's troops but carefully ignores Heth's. Why is this? All soldiers know that the numbered of killed is the one and only test of pluck and endurance. Besides five North Carolina and three Mississippi regiments there were troops from Tennessee, Alabama and Virginia in Heth's division, and all of them behaved gallantly, except the left brigade, which was Brockenborough's Virginia, and its loss in killed in this battle was 25, or five to the regiment, Pickett's 15 (Virginia) regiments had 224, an average of nearly 1t to the regiment. I have never seen the casualties of the Eleventh Mississippi but the numbered of killed in the Second and Forty-Second Mississippi and five North Carolina regiments were 338, an average to the regiment of something over 48. The fire of Cemetery Hill was concentrated upon Heth's division, and at the close of the charge its organization was, to a great extent, broken up; but with the exception of the left brigade, this disorganization was caused by death and wounds.

"Pickett's division, when 1500 of the surrendered in an open field was better organized, but this fact was owing to their comparatively small loss in killed and wounded. The figures are official. I was at Gettysburg and there severely wounded."

A member of the Eleventh Mississippi - but whose name and address I cannot now just now recall - sent me, several years ago, a report of the losses of the Eleventh Mississippi in that battle. He said the regiment went in with 446 privates and 22 officers.

When it came out it had 90 privates and two lieutenants on foot and able to answer roll call next morning. What was left of the regiment was commanded by Lieut. Steve Moore of Aberdeen. This looks like Heth's men did some charging too.

Thomas Dwight Witherspoon was a private in the Lamar Rifles of the 11th Mississippi Infantry. He later served as Chaplain of the 2nd Mississippi and the 42nd Mississippi regiments of infantry.

(H. Grady Howell, Sr.)

"Pickett's Charge" is a romance, and has lived long enough. Fifty years is more than the average life of a newspaper fiction, but this one promises to stay a half century longer – like the old fable of Lee tendering his sword to Grant generously declining to accept it. The only thing that mars the pathetic beauty of the story is there was not a syllable of truth in it.

Vicksburg Evening Post, August 23, 1913.

The following men of the University Greys fought at Gettysburg, July 1-3 according to the official roster and some private records:

Wounded

Baker, A.J. (Captured)	Hicks, Charles A. (Captured)
Cochran, W.H. (Mortally)	Hurt, William J. (Captured)
Dabney, N.J. (Captured)	King, J.J.
DeGraffenreid, W.F.	Lee, Willis Monroe
Estes, W.T. (Mortally)	McKay, J.S.
Goodwin, R.W. (Captured)	McKie, Joseph L. (Captured)
Hall, William R. (Mortally)	McKie, Thomas F. (Mortally)
Harris, Edward L. (Captured)	Moore, John V. (Captured)
Heslep,Almouth	Myers, Calvin R. (Captured)
Heslep, Rufus P.	Tabner, William (Captured)

Killed

Ballard, James E.	Gage, Jeremiah S.
Brewer, Samuel M.	Heslep, Thomas
Dailey, Frank O.	Raines, William A.

* * *

CLARION-LEDGER
FEBRUARY 28, 1877

CHAPLAIN THOMAS D. WITHERSPOON
42ND MISSISSIPPI INFANTRY

Jottings of Army Life.

By Rev. T. D. Witherspoon, D. D. Chaplain of Davis Mississippi Brigade.

Last Days of Col. Hugh R. Miller.

From *Pontotoc Weekly Folio*.

Amongst these recollections, partly bright and partly painful, which give to this period of our history a peculiar charm, none come over us with more mellowed and thoughtful interest, than those which cluster around the form of him whose name stands at the head of this article, whose sands were so soon to be numbered, and of whom it is not too much to say that he was one of the most remarkable men of his day. Those who knew him at home, as an able jurist, a sagacious statesman, an incorruptible patriot, a brave, wise, chivalrous leader of public opinion, will agree with me that few men of his day were more endowed with higher or more ennobling gifts than he.

In the army he had few equals as a Regimental commander. Rigid in the administration of discipline, and yet thoughtful and kind in caring for his men, prompt in the execution of orders, yet always wary and on the alert; doomed to serve under officers whose military abilities were far superior to his own, yet too much of a soldier to refuse obedience, even where he chafed under a sense of the blunders which he was aiding to enact; possessed of a fortitude in the endurance of hardship and privation that nothing could conquer, and a courage on the battlefield that seemed utterly oblivious of all danger, and absolutely vivid of fear.

One or two incidents will serve to illustrate his fortitude and courage. On one occasion whilst camped at Lawrence's Ford, we had a long spell of the most fearfully gloomy weather. The rain dripped incessantly day after day. The water rose in our tents and covered the ground. Our chimney's all refused to draw, and our tents were filled almost to suffocation with smoke. Large fires were kindled out of doors and the tents abandoned. All reading, writing and other indoor work was for the time abandoned. Our man, however, was the constant wonder of the camp. I see him now, sitting in his tent amidst a cloud of shifting smoke, his pen in hand and his regimental rolls spread out over his camp chest, stopping every minute to wipe his glasses free from the moisture the hot smoke was drawing from his inflamed eyes, and yet working on, hour after hour, with an indomitable will that never gave way to anything. And I hear him now, as, upon my remonstrance, he looked up with that quiet but determined air, and said "Work must be done, sir, smoke or no smoke!" It was the secret of his whole success in life, "Work must be done sir," and he was made of that kind of stuff that would die inch by inch before it would shirk duty or shift responsibility.

The second incident will recall itself to the minds of the men as illustrating the courage of Col. Miller on the battlefield.

On the first day of the battle of Gettysburg, Davis' Brigade moved upon one of the strongest positions in the enemy's line of defense. It so occurred that in making the assault, the right and left wings of the 42d Mississippi Regiment fell to the right and left of a railway track, which ran almost exactly in the direction of our advance upon the enemy. The valley on which we were making our advance was a good deal depressed, and consequently, the railroad track elevated on an embankment some twelve or fifteen feet high. The color guard being to the left of the embankment, could not be seen by the men on the right, and there was difficulty in keeping the alignment. Col Miller seeing this, instantly ordered the color sergeant to leave his place in the line and mount the embankment with the colors of the regiment, while the Colonel taking his place immediately under the colors, shouted to the officers on either hand to align their companies on him. I never expected to see a grander sight than that of Col. Miller, his sword at a carry, his tall and stately form backed against the western sky, his elastic step like that of a boy, his conspicuous position making him the target of cannonneers, riflemen and sharpshooters, and yet, as calm and self-possessed amidst the bursting shells, and rattling grapeshot, and whizzing balls, as if he was moving to dress parade.

But it was not of Col. Miller's soldierly qualities I wished to write; but of his religious experience - that of which, since the death of her whom alone he admitted fully to a share of his thoughts and feelings upon religious subjects, and whose womanly devotion to him was so great that his death on the battlefield was soon followed by hers of a broken heart -there is probably no one living who can speak so fully as myself.

From the time when I entered the Chaplaincy of his regiment, he was pleased to admit me to his full confidence and intimate association. For years before entering the army he had been seriously impressed with the necessity of a religious life, and, at times, almost persuaded to give himself up to it.

On several occasions during the earlier months of my chaplaincy he expressed a desire to become a Christian. His bible was with him on all his marches, and not only moments but hours were devoted to its perusal.

Never, however, until after the communion meeting at Ivor Station did he give expression to his determination to devote himself to the life of a Christian. How vividly the whole scene comes up before my mind! We had just broken camp at Ivor and set out upon the open campaign. He realized fully that stern work was before us. He had sent me some weeks before to take Mrs. Miller, who had been with him in camp, to her relatives in Cumberland county to remain until the close of the summer campaign. We were riding together, one of the field officers having kindly lent me his horse for a little while, and I was telling him of my parting with Mrs. Miller, and of her saying as she pressed my hand at parting, "Oh, if my husband were only a Christian." There was a long silence as we rode along. He seemed absorbed in though. At length he said, "There has been a great change in my views and feelings lately. I felt on communion Sabbath as if I really longed to be with you at the table and I hope I shall be at the next, but I am so unworthy" I looked up and there was a tear in my eye, the only one I ever saw there. We conversed together for half an hour and I left him without a doubt upon my mind, that he was a true believer in Christ, though he was without the evidence he craved for himself.

After we had entered Pennsylvania, and only a day or two before the battle of Gettysburg, I was again riding in his company, and he was speaking of the impending battle. Said he, "I am fully convinced that only God's help can gain us this battle. We must look to Him. The odds are all against us." I said, "I hope, Colonel, your personal trust is in Him." "Yes, sir, I believe it is. I sought him as a lost Savior - given myself up to Him, I don't think there is anything I would not do to serve him; but still I feel unworthy."

On, that 3rd day of July! That charge upon Cemetery Hill! My mind becomes bewildered as I think of it! My heart grows sad as I recall the brave and noble ones that bit the dust in that fatal assault! And of them all none was nobler or braver than Hugh R. Miller.

He fell, mortally wounded, while charging at the head of his command and within a stone's throw of the enemy's works. It was impossible to bear him away, and he fell a prisoner into the enemy's hands. There he was found by his son a day or two afterwards, and borne to the house of a kind Christian family who were natives of Maryland and tenderly and skillfully nursed. The wound, which had at first been pronounced fatal, began to heal and surgeons pronounced him out of danger, and his arrangements were all being made to be sent through the lives under special exchange. During this interval of a week or more of convalescence, his mind was, as I learned from pious members of the household where he was entertained, constantly exercised on religious subjects. His son had informed him that I was within the Federal lines in charge of the wounded at the field-hospital, and he had time and again expressed his great anxiety to see me. Not knowing where he was, I had searched through all the hospitals for him in vain, and had given up the hope of seeing him, when a messenger came for me, saying that secondary hemorrhage has ensued, and he was dying from his wound. I hastened to his side to find him already speechless. He asked the attendants again and again if I were not yet come. As I entered the room his face brightened up with a joyous smile, and evidently summoned all his energies to tell me something which he was anxious to communicate, but his lips were already sealed in death. He made a motion with his hand for pencil and paper and they were quickly given to him, but the fingers were too much benumbed to do the office of the lips. A shade of disappointment passed over his face. He had evidently been buoyed up by the hope of communicating with me. Not a moment was to be lost, for the life current was ebbing with every breath. I bent over him and whispered in his ear. "Shall I tell her, Colonel, that her prayers are answered that you died happy, trusting in Christ?" A smile overspread the ashy face, the eyes kindled once more with a glance of intelligence, the lips slightly moved and the head gently nodded assent. And then the eyes closed, and the pulse almost instantaneously ceased to beat, and over the cold clay lingered the smile which the last words he ever heard on earth had awakened.

* * *

Daily Commercial Herald (Vicksburg, MS)
July 19, 1896

Private Dennis Cooney
21st Mississippi Infantry

Succor The Veteran

Somedays ago the C.H. published an appeal to the State authorities from Mr. Dennis Cooney, formerly a resident of this city and member of the 21st Mississippi Regiment, who is now disabled from the effects of a severe wound received at Gettysburg, and who cannot expect even the small relief to be afforded by the State pension, as he is not living in Mississippi. It is learned on most positive authority, that of members of the regiment, that he was in the service and was shot down, as he narrated, at Gettysburg, being the first man hit that day in in the regiment, if not in the brigade. No one of his old comrades who saw him fall supposed that he was alive and to hear from him after the lapse of thirty-two years was a great surprise to them. At the request and that of other sympathizing citizens the C. H. has opened a subscription list for his benefit and will receive a receipt for all money sent to this office for his benefit.

* * *

Macon Beacon
February 5, 1932

Noxubee Rifles, 11th Mississippi, Company F

A Faithful Old Negro

The publication of the roll of the Noxubee Rifles last week following the close of Mr. Mahone's diary, brings to mind on an incident that happened in Macon several years ago, while the late J. F.

Jones was in the drug business here. Mr. Jones came to Macon shortly after the Civil War, having been a member of the above company in the medical corps, Dr. H. A. Minor being the surgeon.

An old negro came into Mr. Jones' store one day and asked Mr. J if he remembered him. Mr. Jones said he could not place him, and the old negro replied, "The first time I saw you was at the Battle of Gettysburg, where my young master, Frank Coleman was wounded. After the battle I went out on the field and found him, and brought him to the camp where you and Dr. Minor were tending to the wounded. My master was badly wounded and died that night, and next day I buried him." The old negro said he 'hoped' Dr. Minor and Mr. Jones cut off the captain's leg the next morning after burying his young master, all of which came back to Mr. Jones recollection.

The old negro's name was Frank Coleman, father of Perry Coleman, and it will always be a lasting shame that he was denied a pension as a servant, to which he was entitled under the law. His last years were spent in poverty on the farms of Mr. Rosette and Mr. Creswell in the Prairie Point neighborhood, who provided for him after he was too feeble to work.

* * *

VICKSBURG JOURNAL
MAY 19, 1866

Gettysburg Dead

To the *News and Mississippian.*

Mr. Editor: The remains of the following named persons are buried upon my farm. If any of their friends or relatives wish to have anything done with them, they can do so by communicating with

J. S. Crawford,
Gettysburg, Adams Co., Penn.

J. W. Snather or Snathell.

J. Adams, Co. A, 13th Miss. Vols.

W. J. Cromer, Co. K. do.

F. M. Ragin. Co. K, do.

N. L. McDuffey, Captain Co. F, 18th Miss. Vols.

H. P. Beauchamp, Co. E. do.

James Worley, Co. F, 21st do.

H. P. Andrew.

Wm. Biggers, Georgia.

N. T. Teader, Co. D, 17th Miss. Vols – His last words were:

"A widow, three sweet babes, fatherless, mourning
Miserre dei Deus."

J. W. Carter, Colonel, 13th do.

R. G. Gunn, Co. A, 17th do.

U. A. Hays, Co. K, 13th do.

J. T. Hallowell, Co. H., do.

J. W. Davis, Co. I, do.

W. B. Connelly, 15th do.

J. C. Rawlson, 13th do.

C. Moore, Co. F. 18th do.

H. Brister, Co. I, 13th do.

J. W. Brice, Co. E, 17th do.

A. P. P. the rest obliterated.

P. A. Robbins, Co. C, 17th do.

J. W. Frazer.

A. C. Reese, Co. F, 10th Regt.

Wm P. Casey,

Wm Layman, 2d, Co. N. A. of N. O.

C. H. Bryan, 3d Co. N. A. of N. O.

P. W. Nell.

Some of the headboards, be deciphered, and many have none at all.

* * *

Semi-Weekly Clarion (Jackson, MS)
September 15, 1871

Mississippi' Dead at Gettysburg.

We find in the *Vicksburg Herald* the following names of Mississippi, who lie at Gettysburg, Pa., and whose graves are marked separately and can be identified. The list is not correct as far as those classed as belonging to the 1st Regiment are concerned. That Regiment was in the Army of Tennessee during the war. The Mississippi Regiments that served in the Army of Northern Virginia, and took part in the battle of Gettysburg, were the 2d, 11th and 42d, in Davis' Brigade; the 12th, 16th, 19th and 48th, in Posey's (afterwards Harris')

Brigade and the 13th, 17th, 18th and 21st in Barksdale's (afterwards Humphrey's) Brigade.

The Herald says: "Dr. R. B. Weaver will be at Gettysburg for several weeks yet, and will disinter any of these remains, pack in boxes and deliver them at boat or cars in Baltimore City, at $4 a piece. "The names are: Of the 1st Regiment, E. P. Hall, Co. I. Of the 2d, J. Leslie; 11th, P. M. Benford, Co. G; Landis Huscksley, Co. E. 12th, J. C. Burley, Co. I, 13th, N. B. Hindman, Co. A; N. A. Parke, Co. K; W. F. Dunken, Co. C; D. Boswell, Co. A; 17th, Lieut L. M. Moore, Co. E; G. F. Co. C; D. Boswell, Col A; 17th, Lieut. L. M. Moore, Co. E; G. F. Box, Co. A; James Mason, Co. I; T. E. Sturgess, Co. F; D. W. McWilliams; Co. B; 18th, H. L. Quinn, Co. E; A. C. Hoffman, Co. A. 19th, T. B. Geoghegan, Co. D. 31st, T. W. Gee, CO. I. 42d, T. J. Anderson, Co. G; John Sansome, Co. D; Lieut. B. F. Wham, Co. B. 48th, Thos Johnson, Co. I; Sergeant F. R. M. Campbell.

* * *

CLARION-LEDGER
AUGUST 15, 1872

Appeal to Patriots

We call the attention of all ex-Confederates to the following appeal, which has been recently issued by ex-Confederates at Richmond, Virginia. Capt. John T. Buck, of this place, has been officially appointed to receive contributions, and, we trust, for the honor of Mississippi, that he will be able to report a considerable amount. Let ex-Confederates in every town and neighborhood collect something, if it be but $5 or $10, and forward it to Capt. Buck. We particularly urge this upon all ex-members of the 2d, 11th, 12th, 13th, 16th, 17th, 18th, 19th, 21st, 42d and 48th Mississippi Regiments, and the friends of those who were members of those regiments. Virginia has already re-interred in Richmond the remains of many Mississippians. Shall she be permitted to do the whole of this work, unaided by Mississippi? Some weeks since we published a list of gallant dead of Mississippi, who at the hands of Virginia have received a final resting place in Hollywood Cemetery.

Appendix A

Confederate Recruitment By State

State	Free Population	Soldiers Recruited	Percentage
Mississippi	354,674	85,000	24.0
Georgia	595,088	130,000	21.8
South Carolina	301,302	60,000	19.9
North Carolina	661,563	127,000	19.2
Florida	78,679	15,000	19.1
Alabama	529,121	100,000	18.9
Louisiana	376,276	53,000	14.1
Virginia	1,105,453	155,000	14.0
Arkansas	324,337	45,000	13.9
Tennessee	834,082	115,000	13.8
Texas	421,649	58,000	13.8

Appendix B
Mississippi Regiments
Order of Battle at Gettysburg[1]

* Muster dates
** Killed at Gettysburg

Barksdale's Brigade
13th MS
Company A: Winston Guards March 16, 1861*
Capt. John M. Bradley
Company B: Wayne Rifles April 17, 1861 Capt. W.J. Eckford
Company C: Kemper Legion April 13, 1861 Capt. James W. Carter**
Company D. Newton Rifles March 23, 1861
Capt. Montgomery Carleton
Company E. Alumutcha Infantry March 23, 1861
Capt. Peter H. Bozeman
Company F. Lauderdale Zouaves March 30, 1861
Capt. Kennon McElroy
Company G. Secessionists April 27, 1861 Capt. Daniel R. McIntosh
Company H. Spartan Band March 23, 1861 Capt. Wesley Millard
Company I. Minute Men of Attala April 19, 1861 Capt. L.D. Fletcher
Company K. Pettus Guards March 30, 1861 Capt. S.J. Randell

17th MS
Company A: Buena Vista Rifles April 22, 1861
Capt. Timothy L. Rogers
Company B: Mississippi Rangers May 8, 1861
Capt. John McGuirk
Company C: Quitman Grays May 23, 1861 Capt. William D. Holder
Company D: Rough and Readies June 7, 1861
Capt. Henry E Williamson
Company E: Burnsville Blues May 11, 1861 Capt. M.D. Morland
Company F: Sam Benton Rifles April 29, 1861
Capt. Bassett G Lawrence
Company G: Confederate Guards April 22, 1861
Capt. W.S. Featherston
Company H: Panola Vindicators April 27, 1861
Capt. George F. Foote
Company I: Pettus Rifles April 23, 1861 Capt. Marmaduke Bell
Company K: Magnolia Guards April 23, 1861 Capt. John W. Lyles

---

1. Stewart Sifakis, *Compendium of the Confederate Armies: Mississippi* (New York, 1995)

18th MS
Company A: Confederate Rifles April 12, 1861
Capt. Joseph M. Jayne
Company B: Benton Rifles April 27, 1861 Capt. William H. Luse
Company C: Confederates April 22, 1861 Capt. O.R. Singleton
Company D. Hamer Rifles April 19, 1861 Capt. Clarence F. Hamer
Company E. Mississippi College Rifles April 20, 1861
Capt. Johnson W. Wellborn
Company F. McClung Rifles April 27, 1861 Capt. Bruce Gerald
Company G. Camden Rifles April 29, 1861 Capt. Adam McWillie
Company H: Brown Rebels April 29, 1861 Capt. Albert G. Brown
Company I. Beauregard Rifles April 18, 1861 Capt. John N. Balfour
Company K. Burt Rifles April 22, 1861 Capt. Erasmus R. Burt

21st MS
Company A: Volunteer Southrons of Vicksburg May, 1861
Capt. Daniel N. Moody
Company B: Transferred to Second Battalion
Company C: Stevens Rifles May 25, 1861 Capt. W.C.F. Brooks
Company D: Jeff Davis Guards May 24, 1861
Capt. William L. Brandon
Company E: Hurricane Rifles June 1, 1861 **Capt. Isaac D. Stamps
Company F: Tallahatchie Rifles June 25, 1861 Capt. W.H. Fitzgerald
Company G: Madison Guards July 1, 1861 Capt. William H. Dudley
Company H: Warren Volunteers June 1, 1861 Capt. Vaughn Noland
Company I: Sunflower Guards August 9, 1861
Capt. B. G. Humphreys
Company K: New Albany Grays June 26, 1861
Capt. Robert W. Flourney
Company L: Vicksburg Confederates August 1, 1861
Capt. Robert L. Green

Davis's Brigade
2nd MS
Company A: Tishomingo Riflemen February 20, 1861
Capt. Bentley B. Boone
Company B: O'Connor Rifles March 4, 1861
Capt. John H. Buchanan
Company C: Town Creek Riflemen February 27, 1861
Capt. William C. Bromley
Company D: Joe Matthews Rifles March 9, 1861
Capt. William D. Beck
Company E: Calhoun Rifles February 8, 1861 Capt. John F. Boothe
Company F: Magnolia Rifles March 4, 1861
Capt. William C. Falkner
Company G: Pontotoc Minute Men March 2, 1861
Capt. Hugh R. Miller

Company H: Conewah Rifles March 1, 1861 Capt. Samuel H. Taylor
Company I: Cherry Creek Rifles April 1, 1861 Capt. John Herring
Company K: Iuka Rifles April 6, 1861 Capt. John M. Stone
Company L: Liberty Guards March 5, 1862 Capt. Robert Storey

11th MS
Company A: University Grays February 7, 1861
Capt. William B. Lowrey
Company B: Coahoma Invincibles February 16, 1861
Capt. Samuel N. Delany
Company C: Prairie Rifles March 2, 1861 Capt. J. Byrd Williams
Company D: Neshoba Rifles April 13, 1861 Capt. A.H. Franklin
Company E: Prairie Guards February 15, 1861
Capt. J.T. Watt Hairston
Company F: Noxubee Rifles February 24, 1861 Capt. George T. Wier
Company G: Lamar Rifles February 23, 1861 Capt. Francis M. Green
Company H: Chickasaw Guards March 19, 1861
Capt. William F. Tucker
Company I: VanDorn Reserve February 20, 1861
Capt. William H. Moore
Company K: Carroll County Rifles February 26, 1861
Capt. Phillip F. Liddell

42nd MS
Company A: Carroll Fencibles May 4, 1862
Capt. Andrew M. Nelson
Company B: Senatobia Invincibles May, 1862
Capt. William A. Feeney
Company C: Nelson's Avengers May ?, 1861 Capt. W.P. Smith
Company D: (Yalobusha County) May 14, 1861
Capt. Robert W. Locke
Company E: Davenport Rifles May 14, 1862
Capt. Henry Davenport**
Company F: (Calhoun County) April 29, 1862
Capt. Thomas G. Clark**
Company G: Gaston Rifles May 13, 1862 **Capt. James H. Gaston
Company H: (Yalobusha County) May 14, 1862 Capt. John Powell
Company I: Mississippi Reds April 21, 1862 Capt. Jeff J. Meek
Company K: (Itawamba County) May 14, 1862
Capt. Goldsborough Mears**

Posey's Brigade
12th MS
Company A: Charlie Clark Rifles March 11, 1861
Capt. John J. McLean
Company B: Natchez Fencibles March 5, 1861
Capt. Edwin M. Blackburn

Company C: Raymond Fencibles March 13, 1861
Capt. William H. Taylor
Company D: Pettus Rifles March 11, 1861 Capt. Merry B. Harris
Company E: Sardis Blues April 20, 1861 Capt. John R. Dickens
Company F: Durant Rifles March 16, 1861 Capt. John A. Carson
Company G: Vicksburg Sharpshooters March13, 1861
Capt. Horace H. Miller
Company H: Claiborne Guards March 25, 1861
Capt. I.G. Hastings
Company I: Satartia Rifles April 24, 1861 Capt. E.R. Gale
Company K: Lawrence Rifles March 30, 1861 Capt. Robert J. Bowen

16th MS

Company A: Summit Rifles April 20, 1861 Capt. J.D. Blincoe
Company B: Westville Guards April 20, 1861
Capt. George D. Funches
Company C: Crystal Springs Southern Rights April 25, 1861
Capt. James C. Davis
Company D: Adams Light Guard May 25, 1861
Capt. Samuel E. Baker
Company E: Quitman Guards April 23, 1861
Capt. Samuel A. Matthews
Company F: Jasper Grays April 27, 1861 Capt. James J. Shannon
Company G: Fairview Rifles March 10, 1861
Capt. John Taylor Moore
Company H: Defenders of Smith County April 29, 1861
Capt. William H. Hardy
Company I: Adams Light Guard May 25, 1861 Capt. Robert Clark
Company K: Wilkinson Rifles April 21, 1861 Capt. Carnot Posey

19th MS

Company A: President Davis Guards April 25, 1861
Capt. J.M. Macon
Company B: Mott Guards May ?, 1861 Capt. W.G. Martin
Company C: Warren Rifles May 8, 1861 Capt. N.H. Harris
Company D: Thomas Hinds Guards May 4, 1861
Capt. Chesley H. Coffee
Company E: McClung Riflemen May 6, 1861 Capt. J.W.C. Smith
Company F: Avant Southrons May 9, 1861 Capt. W.G. Vaughan
Company G: Springport Invincibles May ?, 1861
Capt. R.S. Abernathy
Company H: Salem Cavalry March 9, 1861 Capt. James Hamer
Company I: Marshall Rifles May 25 , 1861 Capt. Thomas J. Hardin
Company K: Jake Thomson Guards May 6, 1861
Capt. W.H.H. Tison

48th MS (formerly 2nd Battalion MS Infantry)
Company A: King Cotton Guards of Vicksburg May, 1861 Capt. Will O. Crutcher
Company B: Rocky Point Rifles May 8, 1861 Capt. T.J. Love
Company C: Dixie Boys August 12, 1861 Capt. Levi C. Lee
Company D: Oktibbeha Rescue July 22, 1861 Capt. J.P. Rogers
Company E: Jackson Boys April 19, 1861 Capt. Richard C. Kerr
Company F: Blewett's Company July, 1861 Capt. Randall Blewett
Company G: Manlove's Company ? ?

(No muster rolls or data regarding any other companies)

BIBLIOGRAPHY

Newspapers

Atlanta (GA) Sunny South
Baltimore (MD) Sun
Benton (LA) Bossier Banner
Brookhaven (MS) Semi-Weekly Reader
Carrollton (MS) Conservative
Charleston (MS) Democratic-Herald
Charleston (MS) Tallahatchie Herald
Clarion (MS) Daily Clarion and Standard
Clarion (MS) Ledger
Columbia (SC) The State
Columbus (MS) Weekly Dispatch
Corinth (MS) Weekly Corinthian
Dublin (TX) Progress
Ellisville (MS) Jones County News
Statesville (NC) Evening Mascot
Fort Worth (TX) Morning Register
Fort Worth (TX) Star-Telegram
Galveston (TX) Daily News
Gettysburg (PA) Compiler
Gettysburg (PA) Star and Sentinel
Greenville (MS) Weekly Democrat-Times
Greenwood (MS) Enterprise
Hernando (MS) Times Promoter
Houston (TX) Post
Jackson (MS) Clarion Ledger
Jackson (MS) Daily News
Jackson (MS) Democratic-Herald
Jackson (MS) Daily Clarion and Standard
Jackson (MS) Vardaman's Weekly
Lake Providence (LA) Banner-Democrat
Lexington (KY) Herald
Lexington (MS) Advertiser
Macon (GA) Beacon
McComb (MS) Enterprise-Journal
McKinney (TX) Daily Courier-Gazette

McKinney (TX) Democrat
Meridian (MS) Daily News
Memphis (TN) Avalanche
Memphis (TN) Commercial Appeal
Miami (FL) Sun-Sentinel
Moulton (AL) Advertiser
Natchez (MS) Weekly Democrat
New Bern (NC) Our Living And Our Dead
New Orleans (LA) Times-Democrat
New Orleans (LA) Times-Picayune
Oxford (NC) Public Ledger
Newton (MS) Herald
Passaic (NJ) Herald-News
Philadelphia (PA) Enquirer
Philadelphia (PA) Weekly Times
Pontotoc (MS) Weekly Folio
Richmond (VA) Dispatch
Richmond (VA) Enquirer
Richmond (VA) Times Dispatch
Savannah (GA) Morning News
Shreveport (LA) Journal
Tupelo (MS) Journal
Vicksburg (MS) Evening Post
Vicksburg (MS) Herald
Wilmington (NC) Messenger
Winona (MS) Times
Yazoo (MS) Herald

Books

Anonymous, *A historical sketch of the Quitman Guards, Company E, Sixteenth Mississippi Regiment, Harris' Brigade: from its organization in Holmesville, 21st April, 1861, to the surrender of the Army of Northern Virginia, 9th April, 1865*, New Orleans, LA: Isaac T. Hinton, 1866.

Ardrey, Joan S., *"A. C." – The Remarkable Life of Alexander Calhoun Ardrey: 21st Mississippi Infantry Army of Northern Virginia 1861-1865*, CreateSpace, 2013.

Ashton, John Douglas. *William Barksdale CSA: A Biography of the United States Congressman and Confederate Brigadier General*, Jefferson, NC: McFarland & Company, 2021.

Bettersworth, John K, (Editor), *Mississippi in the Confederacy: As They Saw It*, New York, NY: Kraus Company,1970.

Bicheno, Hugh, *Gettysburg*, London, UK: Cassell & Co., 2001.

Brasher, Michael T., *The Second Mississippi Infantry Regiment: An Introduction to the Full Annotated Roster*, Lexington, TN: Coonewah Creek Publishing, 2020.

Brown, Maud Morrow, *Lamar Rifles: A History of Company G Eleventh Mississippi Regiment C.S.A., May, 1861 to April, 1865*, Oxford, MS: NP, 1901.

Brown, Maud Morrow. *The University Greys: Company A Eleventh Mississippi Regiment Army of Northern Virginia, 1861-1865*. Richmond, VA: Garret and Massie Inc., 1940.

Christ, Elwood W., *"Over a Wide, Hot ... Crimson Plain": The Struggle for the Bliss Farm at Gettysburg, July 2nd and 3rd 1863*, Baltimore, MD: Butternut & Blue, 1992.

Cockrell, Thomas D., and Ballard, Michael B., (editors,) *A Mississippi Rebel in the Army of Northern Virginia: The Civil War Memoirs Of Private David Holt*. Baton Rouge, LA: Louisiana State Press, 1995.

Coco, Gregory A., *Wasted Valor: The Confederate Dead At Gettysburg*, Gettysburg, PA: Thomas Publications, 1992.

Coddington, Edwin B., *The Gettysburg Campaign: A Study in Command*, New York, NY: Charles Scribner's Sons, 1968.

Cooke, Charles M., Clark, Walter R.(ed,) *History of the Several Regiments and Battalions From North Carolina, 1861-1865*, Goldsboro, NC: Nash Brothers, 1901.

Dawes, Rufus R., *An Officer in the Iron Brigade (1890)*, Bolton, ON: Big Byte Books, 2017.

Dawes, Rufus R. *Service With the Sixth Wisconsin Volunteers*. Marietta, OH: E.R. Alderman & Sons, 1890.

Evans, Clement A. (ed.), *Confederate Military History*, 12 Vols., Atlanta, Ga: Confederate Publishing Company, 1889.

Evans, Robert G. (ed), *The 16th Mississippi Infantry: Civil War Letters and Reminiscences*, Jackson, MS: University Press of Mississippi, 2002.

Gallagher Gary W.(ed.), A.J.L. Freemantle., *"The Gettysburg Campaign," Two Witnesses at Gettysburg: The Personal Accounts of Whitelaw Reid and A.J.L. Freemantle*, St. James, NY: Brandywine Press,1994.

Gallagher, Gary W. (ed.), *Lee: The Soldier*, Lincoln, NB: University of Nebraska Press, 1987.

Gallagher, Gary W. (ed.), E. P. Alexander, *Fighting for the Confederacy*, Chapel Hill. NC: University of North Carolina Press, 1989.

Giambrone, Jeff T., *Remembering Mississippi's Confederates*, Charleston, SC: Arcadia Publishing, 2012.

Giles, S. H., *War Years C.S.A.: 12th Mississippi Infantry Regiment: Major S. H. Giles, Q.M.: Original Letters1861-1865*. Hillsboro, TX: Hill Jr. College Press, 1998.

Gottfried, Bradley M., *Brigades Of Gettysburg: The Union and Confederate Brigades at the Battle of Gettysburg*, New York, NY: Skyhorse Publishing Inc., 2012.

Hankins, Samuel W., *Simple Story of a Solider: Life and Service in the 2d Mississippi Infantry (1912)*, Tuscaloosa, AL: The University Of Alabama Press, 2004.

Heflin, W. P., *Blindman on the Warpath*, Philadelphia, MS: Book, 1903.

Hessler, James A., & Isenberg, Britt C., *Gettysburg Peach Orchard: Longstreet, Sickles and the Bloody Fight for the Commanding Ground Along the Emmitsburg Road*, Eldorado Hills, CA: Savas-Beatie, 2019.

Hoke, Jacob, *The Great Invasion of 1863* (1887), Dayton, OH: W.J. Shuey Publisher, 1887.

Howell, H. Grady, Jr, *For Dixie I'll Take My Stand: A Muster Listing of All Known Mississippi Confederate Soldiers, Sailors, Marines*, 4 Vols., Madison, MS: Chickasaw Bayou Press, 1998.

Hoyle, Joseph J., *Deliver Us From This Civil War: The Civil War Letters of Lieutenant Joseph J. Hoyle, 55th North Carolina Infantry*, Jefferson, NC: McFarland & Company, 2010.

Ladd, David L. and Audrey J. (ed.), *The Bachelder Papers: Gettysburg in Their Own Words*, 3 Vols., Dayton, OH: Morningside House, Inc. 1995.

Laino, Philip, *Gettysburg Campaign Atlas*, Trumbull, CT: Gettysburg Publishing LLC., 2015.

Lightsey, Ada Christine, (ed.) *The Veteran's Story*, Meridian, MS: Meridian News, 1899.

Martin, James M., et al., *History of the Fifty-Seventh Pennsylvania Volunteer Infantry*, Kearny, NJ: Belle Grove Publishing Co. Reprint, 1995.

Mclean, Jess, N. Sr., *The Official Records of the 13th Mississippi Infantry Regiment: as told by those who were there*, Dallas, TX: Published Independently, 2001.

Muffly, J. W. (ed,) *The Story of Our Regiment: A History Of The 148th Pennsylvania Vols., Written By The Comrades*, Des Moines, IA: Kenyon Printing & Mfg. Co., 1904.

Murray, Robert L., *The Redemption of the Harper's Ferry Cowards*, Wolcott, NY: Benedum Books, 1994.

Owen, Joe, *Lone Star Valor: Texans of the Blue and Gray at Gettysburg*, Trumbull, CT: Gettysburg Publishing LLC, 2019.

Pfanz, Harry W., *Gettysburg: The Second Day*, Chapel Hill, NC: University of North Carolina Press,1987.

Robertson, James I. Jr., *General A.P. Hill: The Story of a Confederate Warrior*, New York, NY: Random House, 1987.

Rowland, Dunbar (ed.) *Publications of the Mississippi Historical Society Centenary Series*, Vol. 2, Jackson, MS. Madison, MS: Democrat Printing Company, 1918.

Rowland, Dunbar, *Military History of Mississippi 1803-1898*, Grady Howell, ed., Madison, MS: Chickasaw Bayou Press, 2003.

Sears, Stephen W., *Gettysburg*, Boston, MA: Mariner Books, 2003.

Silver, James W., (ed.) *A Life For the Confederacy: From the War Diary of Robert A. Moore, Pvt., C.S.A.*, Jackson, TN: McCowat-Mercer Press Inc., 1959.

Sorrell, Brig. *Gen. G. Moxley CSA, Recollections of a Confederate Staff Officer*, Bell Irvin Wiley (ed.), 1959, Wilmington, NC: Broadfoot Publishing Reprint, 1995.

Stackpole, Edward J., *They Met At Gettysburg*, Harrisburg, PA: Eagle Books, 1956.

Stubbs, Steven H., *Duty-Honor-Valor: The Story of the Eleventh Mississippi Infantry Regiment*, Brandon, MS: Quail Ridge Press, 2000.

Sutton, E. H., *Grand Pa's War Stories*, Demorest, GA: Banner Printing Company, 1910.

Wilds, Ellen Sheffield., (ed,) *Far From Home: The Diary of Lt. William H. Peel 1863-1865*, Bloomington, IN: Author House, 2009.

War of the Rebellion: A Compilation of the Official Records of the Union and Confederate Armies, 128 Vols., Washington, DC: Government Printing Office, 1880-1901.

Williams, T. P., *The Mississippi Brigade of Brig, Gen. Joseph R. Davis: A Geographical Account of Its Campaigns and a Biographical Account of Its Personalities, 1861-1865*, Dayton, OH: Morningside Bookshop, 1999.

Willson, A.M., *Disaster, Struggle, Triumph. The Adventure of "1000 Boys In Blue," From August, 1862 to June 1865*, Albany, NY: Argus Company Printers, 1870.

Wilson, Legrand James, *Memoir, 1902, The Confederate Soldier*, Reprint, Bell Wiley (ed.), Memphis, TN: Memphis State University Press, 1973.

Yeary, Mamie, *Reminiscences of the Boys in Gray 1861-1865*, Nacogdoches, TX: Ericson Books, 2003.

Young, Jesse B., *The Battle of Gettysburg: A Comprehensive Narrative*, New York, NY: Harper & Brothers Publishers, 1913.

Periodicals

Alexander, E.P., "Artillery Fighting At Gettysburg," *Battles & Leaders of the Civil War*, Vol. 3, New York, NY: Castle Books, 1956.

Belo, A.H., "The Battle of Gettysburg," *Confederate Veteran*, vol. 8, No. 4, April 1900.

Campbell, Eric A., *"Sacrificed to the bad management ... of others": Richard H. Anderson's Division at the Battle of Gettysburg, The Army of Northern Virginia in the Gettysburg Campaign 1998*, Gettysburg National Military Park Seminar 1998, Gettysburg, PA: GNMP, 1999

Campbell, Eric A., *"The Key to the Entire Situation—The Peach Orchard, July 2, 1863," The Second Day at Gettysburg*, 2006 Gettysburg National Military Park Seminar, Gettysburg, PA: GNMP, 2008.

Clark, George, "Wilcox's Alabama Brigade At Gettysburg," *Confederate Veteran*, Vol. 17, No. 5. May 1909.

Cummings, C. C., "Gettysburg, Romance In Prison," *Confederate Veteran*, vol. 23, No. 4. April 1915.

Cummings, C. C., "Chester Hospital On The Delaware," *Confederate Veteran*, April 1916, Vol. 24, No. 4.

Duke, J. W., "Mississippians At Gettysburg," *Confederate Veteran*, Vol. 14, No. 5, May 1906.

Harris, T. C. "The Second Mississippi At Gettysburg," *Confederate Veteran*, Vol. 25, No. 11, November 1917.

Hood, Maj. John J., "Tribute to Gen. Barksdale," *Confederate Veteran*, November, 1901; "Correction," *Confederate Veteran*, February, 1902.

Hoopes, John (ed,) "The Confederate Memoir of William M. Abernathy," *Confederate Veteran*, Vol. 3, 2003.

Howell, Frank A., "Eleventh Mississippi A Gettysburg," *Confederate Veteran*, Vol. 19, no. 7, July 1911.

Johnson, W. Gart, "Reminiscences Of Lee And Of Gettysburg," *Confederate Veteran*, Vol. 1, No. 8, August 1893.

Love, William A., "Forward and Back," *Confederate Veteran*, Vol. 33, No. 1, January 1925.

Love, William A., "Mississippi At Gettysburg." *Publications of The Mississippi Historical Society*, Vol. IX, 1906.

McFarland, Baxter, "Casualties of the Eleventh Mississippi Infantry Regiment at Gettysburg," *Confederate Veteran*, Vol. 14 No. 3, March 1919.

McLaws, Major General Lafayette, "Gettysburg," *Southern Historical Society Papers*, Vol. 7., January 1879.

McNeily, J. S., "Barksdale's Mississippi Brigade at Gettysburg," *Publications of the Mississippi Historical Society*. Vol. XIV, 1914.

Mississippi in the War Between the States: Civil War Centennial, Jackson, MS: Mississippi Commission On The War Between The States, 1960.

Reid, W. D., "By W. D. Reid, Houston, Miss." *Confederate Veteran*, Vol. 19, No. 2, February 1911.

Strain, J. H., "Heroic McPherson," *Confederate Veteran*, Vol. 31, No. 6., June 1923.

"The Civil War Almanac: A Book of Compelling Facts, Figures, and Expert Opinions," *Civil War Monitor*, Kettering, Ohio: January 2018.

Watts, A. T., "Something More About Gettysburg," *Confederate Veteran*, Vol. 6, No. 2, February 1898.

Youngblood, William., 'Unwritten History of the Gettysburg Campaign," *Southern Historical Society Papers*, Vol. 38, 1910.

Diaries/Letters/Manuscripts

Diary, Sgt. William Hill, 1861-1863, Mississippi Department of Archives and History, Z0217.00

Diary, Robert A. Moore, Mississippi Department of Archives and History, https://egrove.olemiss.edu/ciwar_diaries/3/.

Diary, Augustus Vairin, Mississippi Department of Archives and History

Letter, Richard C. Bridges to his sister Carrie, July 22, 1863. Richard C. Bridges Collection, Special Collections, University of Mississippi Libraries.

Letter, H. Q. Bridges to Mrs. V. G. Armistead, July 26, 1913. Gave Family Collection, Special Collections, University of Mississippi Archives.

Letter, Jeremiah Gage to Patience Gage, July 3, 1863, Gage Family Collection, Special Collections, University of Mississippi Libraries.

Holt, David Eldred. Ms. Mississippi Department of Archives and History, n.d.

Humphreys, Benjamin Grubb, Unpublished Autobiography, Mississippi Department of Archives and History

Letter, Nelson, William Cowper, "William C. Nelson to J. H. Nelson to J. H. Nelson (July 1863). Correspondence 690. https://egrove.olemiss edu/ciwar_corresp/690. Accessed July 11, 2020.

Letter Nelson, William Cowper, "William C. Nelson to Maria C. Nelson (19 July 1863).

Correspondence. 691. https://egoveolemiss.edu/ciwar_corresp/691.

Letter, McLaws, Lafayette B.G. Humphreys to McLaws, January 6, 1878, in McLaws Papers, Southern Historical Collection, University of North Carolina Archives.

Nash, Nimrod Newton letters, Old Courthouse Museum Collections, Vicksburg, MS.

Lectures/Videos

Jeff Giambrone, "An Illinois Rebel at Gettysburg." YouTube, April 2022. Youtube.com.watch?V-Wu1t VXII.Tec&t=1165s. Accessed November 15, 2022.

Webpages

Andrew Brown (ed.), "2nd Mississippi Regiment Infantry, Old Ord's War Journal." Accessed June 30, 2020. https://sites.rootsweb.com/ ~ mscivilw/vairindiary.htm. Accessed December 7, 2022.

Elmore, Tim, E. Howard 2015, January 16, E. Howard, McCaleb, in Civil War Talk,

https://civilwartalk.com/threads/sergeant-major-e-howard-mccaleb-jr-of-the-12th-mississippi.155310/. Accessed January 15, 2023.

Tom Howard, "Sergeant Major E. Howard Macaleb of the 12th Mississippi," in Civil War Talk, February 27, 2019. https://civilwartalk.com/threads/sergeant-major-e-howard-mccaleb-jr-of-the-12th-mississippi.155310/. Accessed December 4, 2022.

Index

Military Units

Confederate Units

About Joseph L. Owen

Joe Owen is a National Park Ranger at Lyndon B. Johnson National Historical Park in Johnson City, Texas. Having served 15 years in the U.S, Navy, he attended college at East Central University in Ada, Oklahoma receiving a Bachelor's Degree in History and a Master's Degree in Secondary Education. He taught Social Studies for eight years in Oklahoma, Texas and Oregon before working for the National Park Service.

He is an avid researcher who has co-authored several books and authored A Fine Introduction to Battle: Hood's Texas Brigade at The Battle of Eltham's Landing, May 7, 1862. He received the Jefferson Davis Gold Medal from the United Daughters of the Confederacy for outstanding research and writing in 2019.

Joe lives in Blanco, Texas.

About J. Douglas Ashton

Doug Ashton is a Canadian retired steel and materials testing executive and a lifelong student of the American Civil War dating to the Centennial in 1961 when he was a young boy. He first became interested in William Barksdale and his Mississippi Brigade upon joining a joint American-Canadian reenactment group in 1983 — Company B, Benton Rifles, 18th Mississippi Infantry. Following approximately thirty years of dedicated research encompassing numerous archives, museums, and battlefields Doug authored the first definitive biography of this controversial politician and fearless Mississippian warrior — *WILLIAM BARKSDALE, CSA: A Biography of the United States Congressman & Confederate Brigadier General* published in 2021.

Doug has also authored several articles for *North & South Magazine* and a feature article for *Gettysburg Magazine*. He resides with his wife, Susie, in the quiet hamlet of Port Rowan, Ontario nestled along the north shore of Lake Erie.